OCR Electronics For A2

OCR
Electronics
For A2

Michael Brimicombe

Orders: please contact Bookpoint Ltd, 130 Milton Park, Abingdon, Oxon OX14 4SB. Telephone: +44 (0)1235 827720. Fax: +44 (0)1235 400454. Lines are open from 9.00 to 5.00, Monday to Saturday, with a 24-hour message-answering service. You can also order through our website www.hoddereducation.co.uk

If you have any comments to make about this, or any of our other titles, please send them to educationenquiries@hodder.co.uk

British Library Cataloguing in Publication Data
A catalogue record for this title is available from the British Library

ISBN: 978 0 340 973 677

First Edition Published 2009
Impression number 10 9 8 7 6 5 4 3 2
Year 2012, 2011, 2010, 2009

Cover photo © Benjamin Shearn/Taxi/Getty Images
Typeset by Servis Filmsetting Ltd, Stockport, Cheshire
Printed in Great Britain for Hodder Education, an Hachette UK Company, 338 Euston Road, London NW1 3BH
by Martins the Printers, Berwick upon Tweed.

The author and publisher would like to thank the following for use of photographs in this volume: Fig 6.1 © Ian Poole/iStockphoto.com; Fig 6.2 © Ian Poole/iStockphoto.com; Fig 6.4 © Candi Van Meveren/iStockphoto.com.

All illustrations by Servis Filmsetting Ltd.

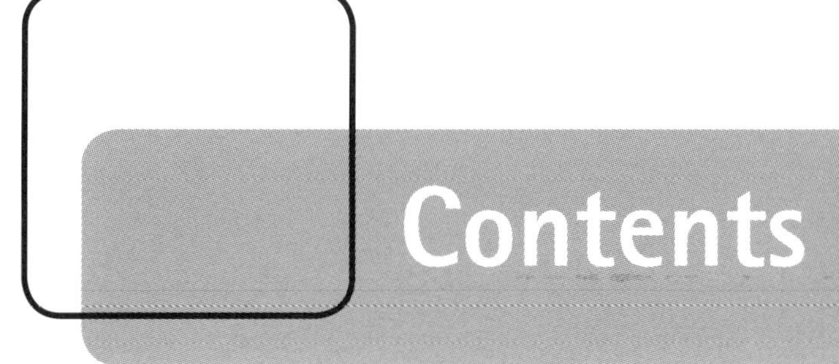

Contents

Introduction

Having survived the AS course, you are now in a position to further your understanding of electronics. The AS course covered some of the devices and ideas which underpin the whole of electronics, and the A2 course gives you an in-depth experience of two important branches of the subject: control and communications.

This book builds on your understanding of AS electronics, showing how these small familiar sub-systems can be combined to make large useful systems. These include memories, microcontrollers, power supplies, radio receivers and the internet. The devices and techniques that you learnt about at AS will be used over and over again as you explore the ways in which they are applied in the real world of computer communications, device control, data storage and information transmission. It may therefore help to have a copy of OCR Electronics for AS to hand as you study this book.

As with OCR Electronics for AS, the book is only part of the course. The other parts can be found at www. hodderplus.co.uk/ocrelectronics. It contains

- exercises to consolidate your understanding
- full instructions for practical work
- data sheets on integrated circuits
- answers to the questions in this book
- guidance for teachers and technicians

There are two ways of working your way through the course. You can, of course, start at the beginning and work through to the end. On the other hand, you could start at Chapter Five and work through to the end before returning to Chapter One. Either way, you should do the practical work and exercises for each section before attempting the exam-style questions. You can only acquire the full flavour of the art which is modern electronics with the correct mixture of theory, practical and self-evaluation.

CHAPTER 1

MOSFET circuits

1.1 Variable resistors

You already know how to use a MOSFET as a driver, as shown in Fig 1.1.

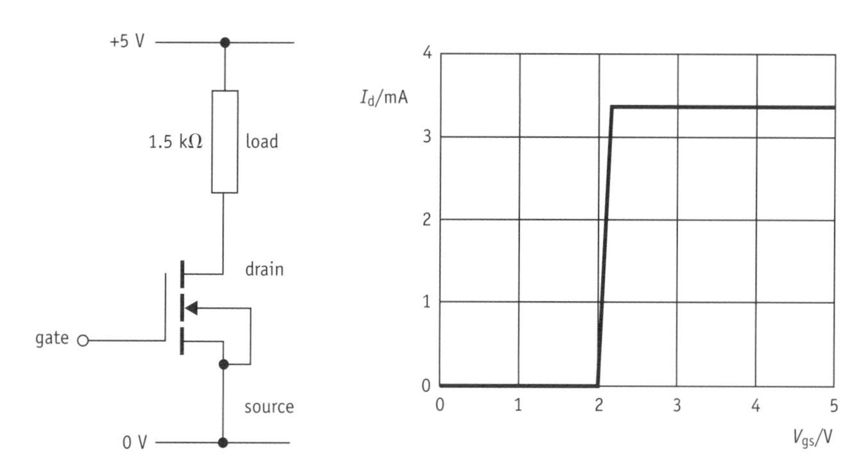

Fig 1.1 The drain current saturates as the gate–source voltage rises above the threshold value

The graph of Fig 1.1 shows how the drain current I_d depends on the gate–source voltage V_{gs}. As V_{gs} rises from 0 V, there is no drain current until V_{gs} reaches the **threshold voltage** V_{th} of 2.0 V. The current then rises very rapidly until it saturates at a value fixed by the resistance of the load.

$I = ?$

$V = 5 - 0 = 5\ \text{V}$

$R = 1.5\ \text{k}\Omega = 1.5\times10^3\ \Omega$

$$R = \frac{V}{I}$$

$$I = \frac{V}{R} = \frac{5}{1.5 \times 10^3} = 3.3 \times 10^{-3}\ \text{A or } 3.3\ \text{mA}$$

The calculation assumes that the drain–source resistance of the MOSFET is effectively zero when it has been switched on. This is reasonable for a MOSFET which has been specifically designed for use as a driver.

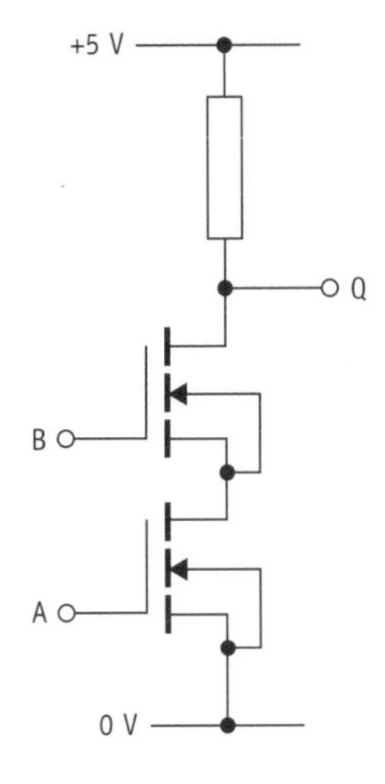

Fig 1.2 The output only goes low when both inputs are high

Gates and op-amps

MOSFETs can do much more than just switch currents in a load on and off. Figs 1.2 and 1.3 show that they can be used to make devices which behave like NAND gates and op-amps.

Since op-amps and NAND gates are the fundamental building blocks of larger systems, such as microcontrollers and mixers, this means that MOSFETs can be the key components underlying the whole of electronics. This versatility comes from the fact that a MOSFET is two devices in one. It can be a voltage-controlled resistor or a voltage-controlled current sink, depending on the voltage between drain and source.

MOSFET characteristic

The graph of Fig 1.4 shows how the drain current I_d of a typical MOSFET depends on the drain–source voltage V_{ds} for different values of the gate–source voltage V_{gs}. The threshold voltage for this particular MOSFET is 1.7 V.

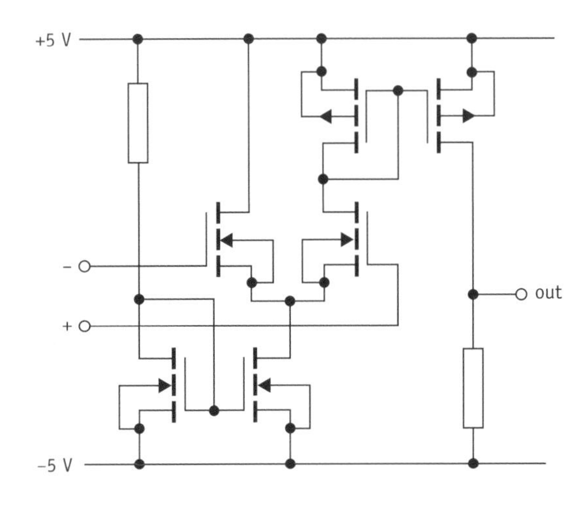

Fig 1.3 An op-amp circuit made from MOSFETs

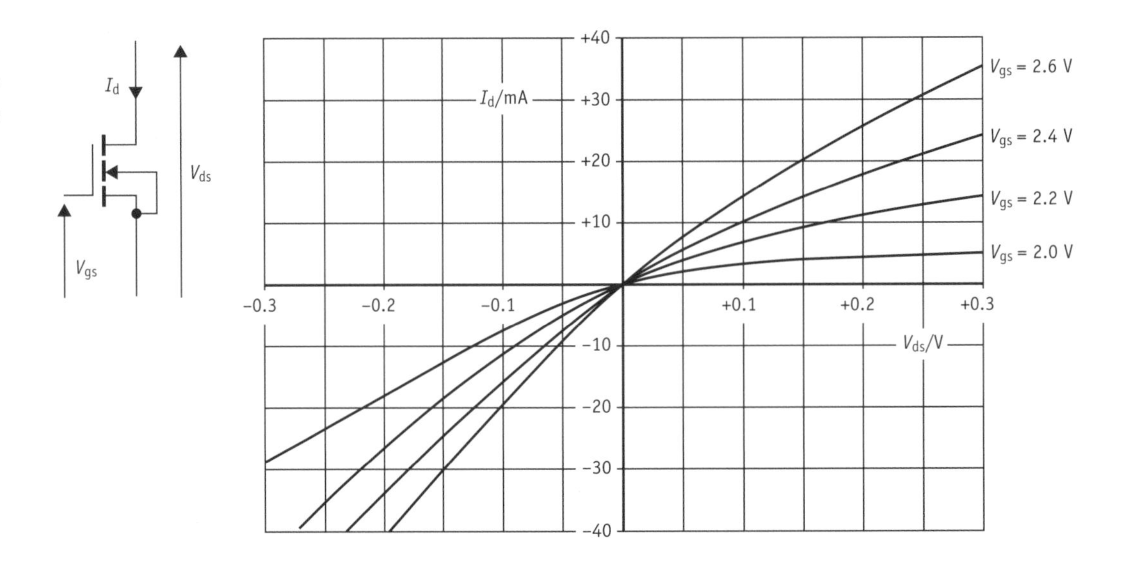

Fig 1.4 MOSFET characteristic: each curve is for a different voltage at the gate

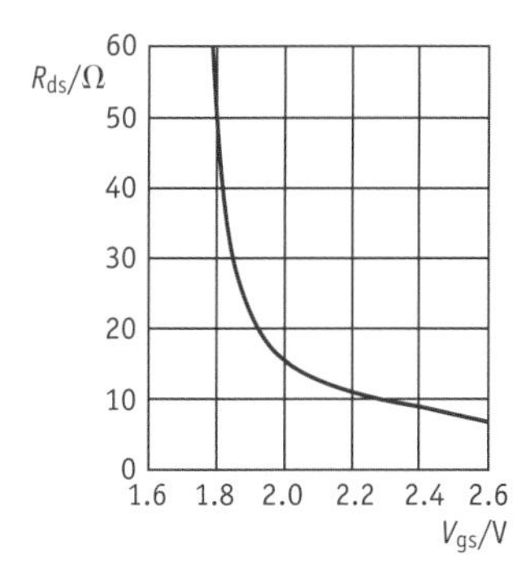

Fig 1.5 The resistance of the MOSFET depends on the gate–source voltage

The four different values of V_{gs} result in curves of the same shape, but with different gradients as they all pass through the origin. The fact that each $I_d - V_{ds}$ graph is a line through the origin suggests that the drain–source resistance R_{ds} of the MOSFET is fixed by the value of V_{gs}. This can only be true for sufficently low values of V_{ds}, typically between +100 mV and −100 mV. The graph of Fig 1.5 shows how the value of R_{ds} of the MOSFET depends on the value of V_{gs} when V_{ds} is between +100 mV and −100 mV. This part of the characteristic is known as the **resistive region**. As you will find out later in this chapter, values of V_{ds} above the threshold voltage result in completely different behaviour.

Potentiometer

The circuit of Fig 1.6 shows how a MOSFET can be used to make a voltage-controlled volume control, or potentiometer. The setting of the wiper is effectively done by altering the value of V_{gs} rather than altering the angle of the shaft of a volume control by hand. Of course, this only works for a.c. signals at **in** whose amplitude is less than 100 mV, keeping the MOSFET well within its resistive region.

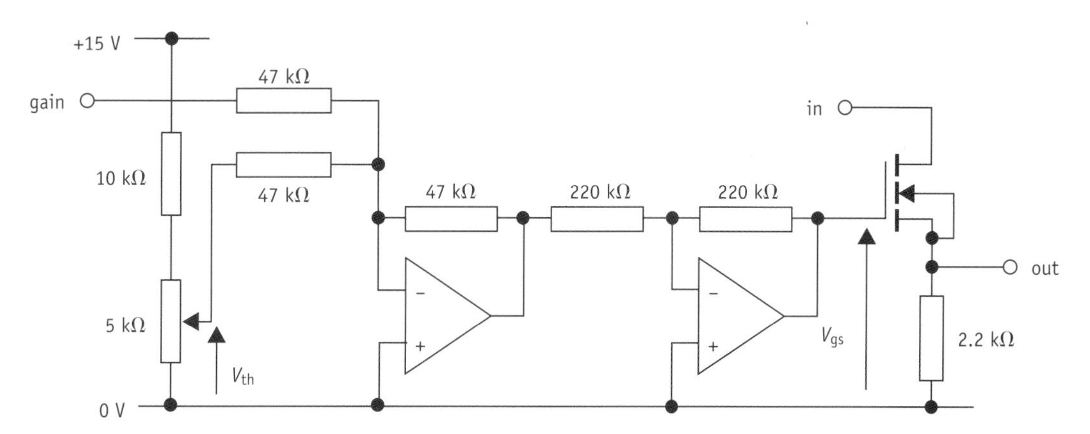

Fig 1.6 A voltage-controlled potentiometer

Note the use of summing and inverting amplifiers to generate the value of V_{gs} which controls the MOSFET. The 5 kΩ potentiometer is adjusted so that its wiper is at V_{th} for the particular MOSFET being used – and each MOSFET will have its own value of V_{th}. This allows a voltage of 0 V at the **gain** input to result in an overall gain of zero, as R_{ds} will then be much bigger than 2.2 kΩ. Increasing the voltage at **gain** raises the value of V_{gs} above V_{th}, reducing the value of R_{ds} and letting more of the a.c. signal at **in** through to **out**.

Analogue switches

An analogue switch is a three terminal device which acts as a voltage-controlled switch. A sufficiently high voltage at the **control** terminal reduces the resistance between the two **in/out** terminals from a very large to a very small value. Fig 1.7 shows how one can be made from a single MOSFET.

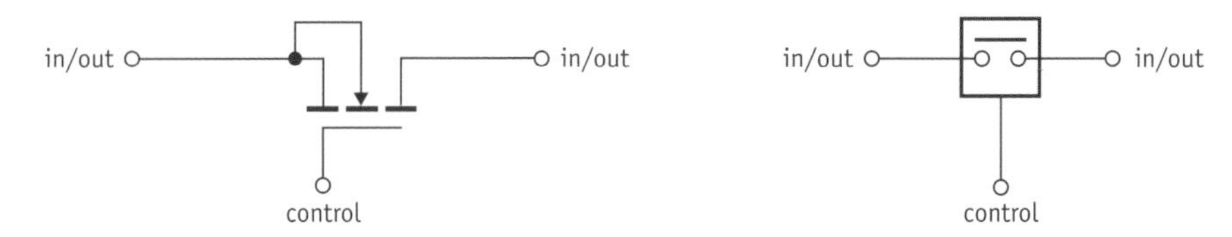

Fig 1.7 A MOSFET can act as an analogue switch

So how does it work? Assume that the MOSFET has a threshold voltage of 2.5 V. Suppose that the gate is held at 0 V. Then provided that neither drain nor the source are below −2.5 V the MOSFET will be in its high resistance state. Now hold the gate at +5 V. This puts the MOSFET into its low resistance state, provided that neither drain or source is above +2.5 V. The overall behaviour is that of a switch that can be turned on and off by a voltage instead of a finger. Of course, the switch can only deal with signals between +2.5 V and −2.5 V, these limits being set by the threshold voltage of the MOSFET.

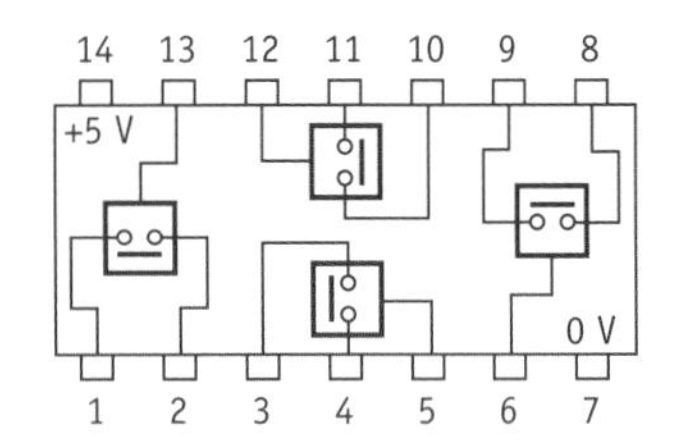

Fig 1.8 The 4066 integrated circuit has four separate analogue switches

The 4066 i.c. has four separate analogue switches, as shown in Fig 1.8. By using parallel pairs of different types of MOSFET, the switch low-resistance value stays at about 100 Ω for signals at **in/out** anywhere between the supply rails. So if you run the i.c. off supply rails at +5 V and −5 V (instead of 0 V), each switch can block or transmit audio signals of amplitude up to 5 V. (Of course, if you do this, the control signals need to be at +5 V to close a switch and −5 V to open it.)

Analogue multiplexers

Two analogue switches and a NOT gate allow you to make a two-bit multiplexer which can handle the two-way flow of audio signals. Depending on the state of A, either S_1 or S_0 is connected to Q, as shown in the table.

control signal at A	switch state
0	S_0 connected to Q
1	S_1 connected to Q

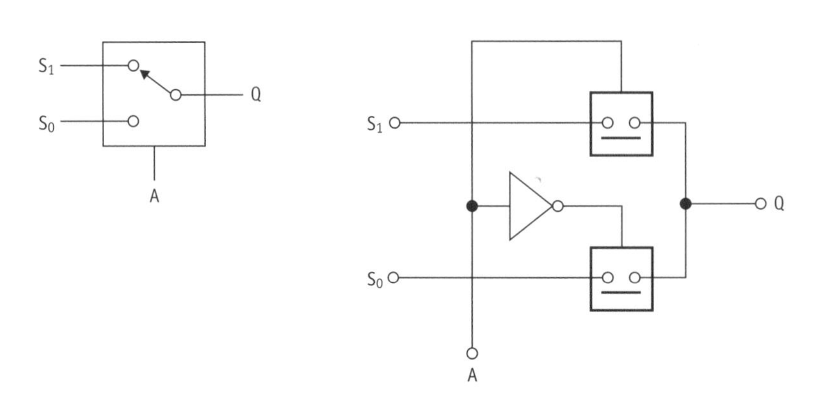

Fig 1.9 A two-bit analogue multiplexer

Routing signals

Because of their two-way nature, analogue multiplexers are also demultiplexers. Furthermore, you can stack them together to make larger versions of themselves. This allows them to be used as routers for audio signals, allowing digital signals to control the flow of analogue signals along wire links, as shown in Fig 1.10.

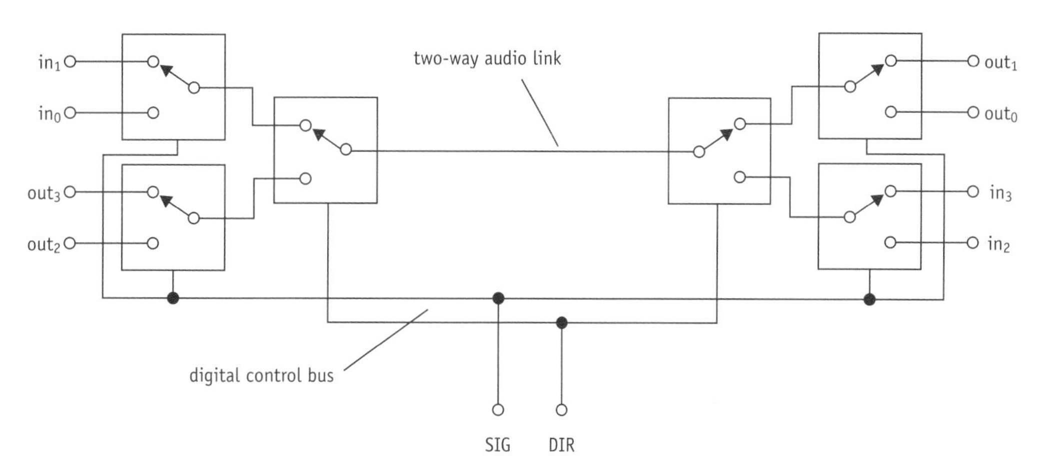

Fig 1.10 Analogue multiplexers allow the two-way flow of audio signals

At any one time, only one of the terminals on the left is connected to just one terminal on the right. The connection is fixed by the pair of digital signals SIG and DIR on the **control bus**. The state of DIR controls the direction of the flow of information down the **audio link**.

DIR	flow direction
0	right to left
1	left to right

The other control signal SIG selects the audio input and output terminals.

DIR	SIG	input terminal	output terminal
0	0	in_2	out_2
0	1	in_3	out_3
1	0	in_0	out_0
1	1	in_1	out_1

Note that because the multiplexers controlled by DIR are made from analogue switches, audio signals can flow both ways through them.

Digital links

The system shown in Fig 1.11 allows a single bit of information to be transferred from A to Y or B to Z along a single link. The direction of transfer is fixed by the digital signal DIR.

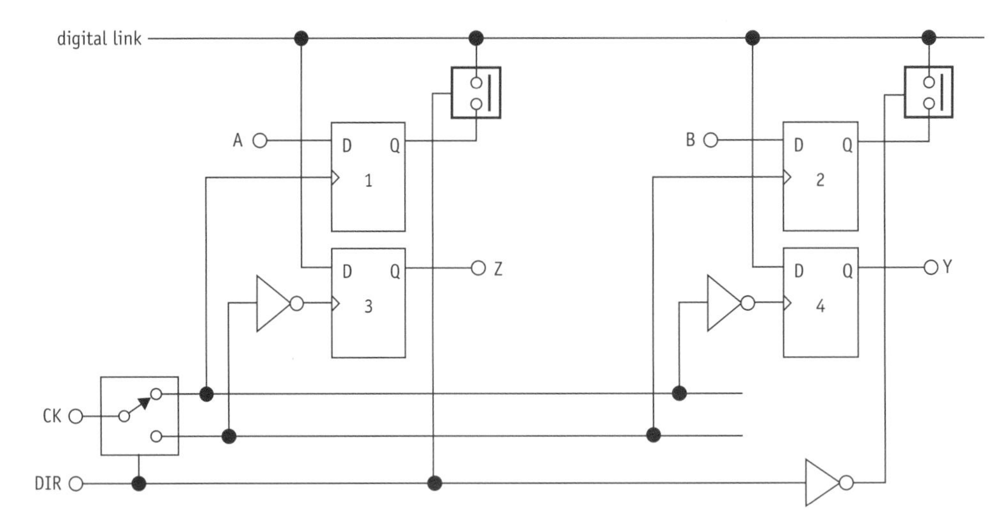

Fig 1.11 Two analogue switches control the flow of bits along the digital link

DIR	flow direction
0	B to Z
1	A to Y

Suppose that you want to transmit a bit from B to Z. Here is the procedure.

- Hold DIR low. This closes the right-hand analogue switch, connecting the output of flip-flop 2 to the digital link. It also routes CK to the clock inputs of flip-flops 2 and 3.
- Push CK high. This causes the signal at the D input of flip-flop 2 to be latched at Q and passed through the closed switch to the digital link.
- Pull CK low again. This forces flip-flop 3 to latch the bit on the digital link and output it at Z.

The procedure transfers a bit from right to left along the link. Doing the same thing, but with DIR held high transfers a bit in the opposite direction.

Notice how the state of DIR only allows one of the flip-flop outputs to be connected to the digital link at any one time, avoiding clashes. It also determines which pair of flip-flops (1 and 4 or 2 and 3) can receive clock signals from CK via the two-bit digital demultiplexer. (You don't have to use analogue switches to isolate the flip-flop inputs from the digital link, as they draw hardly any current from it.)

Tristates

The analogue switches in Fig 1.11 only transfer information in one direction, from a flip-flop output to the digital link. This means that they can be replaced by **tristates**, three-terminal devices which control the one-way flow of digital signals.

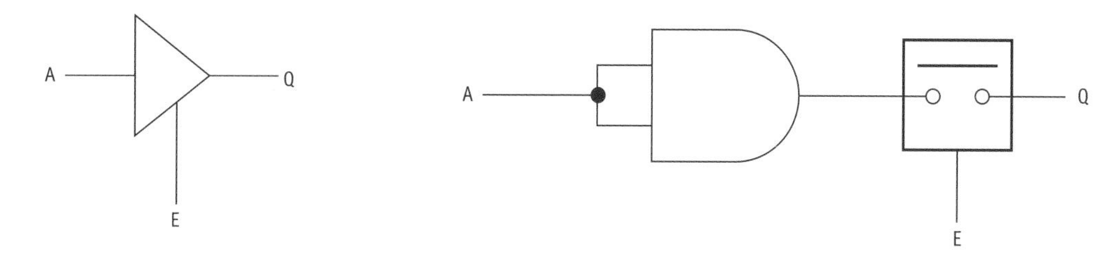

Fig 1.12 The output of a tristate is not connected to anything when E is low

The circuit symbol for a tristate is shown on the left-hand side of Fig 1.12.

The output terminal Q has the same state as the input A when the control terminal E is high, enabling the tristate. However, when E is low, Q **floats** – it behaves as though it weren't connected to anything at all. The '–' symbol in the truth table indicates **open circuit**.

E	A	Q
0	0	–
0	1	–
1	0	0
1	1	1

The circuit on the right-hand side of Fig 1.12 shows one way of implementing a tristate. When E is high, the analogue switch is closed, allowing the state of A to be fed through to Q. Opening the switch by holding E low leaves Q effectively connected to nothing at all – its voltage isn't determined by the signal at A anymore.

Tristates are widely employed in microprocessor systems where bits need to be transferred from one sub-system to another along digital links, as shown in Fig 1.13. The right-hand flip-flop can latch data from the digital link. That data can come from the output of any one of the three flip-flops on the left, provided that the appropriate tristate has been enabled.

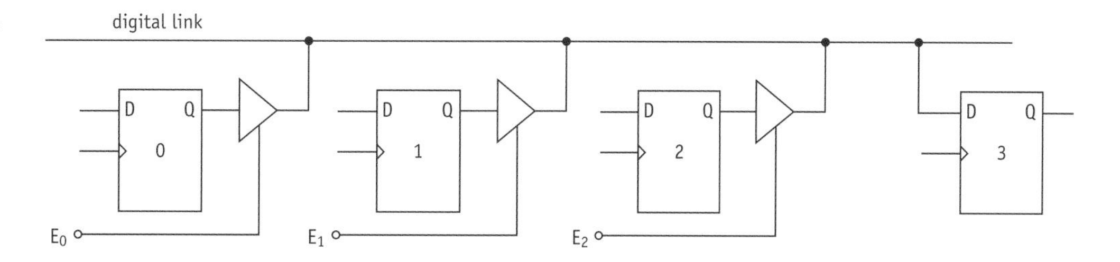

Fig 1.13 Tristates allow more than one flip-flop to place a signal on the digital link

1.2 Amplifiers

It can be argued that electronics as a separate discipline was born in 1911. That was when a three-terminal device called a thermionic valve was first used to increase the amplitude of an a.c. signal. Such devices underly all the important building blocks of electronics, including logic gates and op-amps. Varying the voltage at one terminal of a valve (the grid) resulted in variation of the current between the other two (the anode and cathode). Valves are large and expensive and they wear out. Over the past hundred years, different technologies have been invented to produce the same effect. The MOSFET is the latest version – small, cheap and long-lived, but it has the same behaviour. Changing the voltage of one terminal still alters the current between the other two.

Current sinks

The graph of Fig 1.14 shows how the drain current I_d of a typical MOSFET depends on the drain–source voltage V_{ds} for three different values of the gate–source voltage.

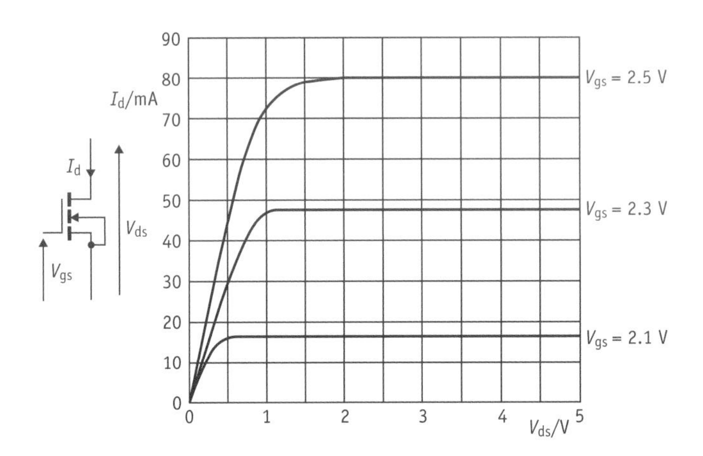

Fig 1.14 MOSFET characteristic for large values of the drain–source voltage

You have already seen the region near the origin in Fig 1.14. This is where the MOSFET behaves like a voltage-controlled resistor. Far away from the origin, the curves become almost horizontal, with the drain current largely independent of the drain–source voltage. This is the **current sink** region of the MOSFET characteristic, where the drain current only depends on the gate–source voltage. Notice how the value of the drain current in this region increases with increasing gate–source voltage. This is summarised in the graph of Fig 1.15.

The drain current I_d of the MOSEFT as a current sink can be modelled quite well with this formula.

$$I_d = g_m (V_{gs} - V_{th})$$

V_{gs} is the gate–source voltage, V_{th} is the threshold voltage and g_m is the **transconductance** in siemens (S). The formula only works well when V_{ds} is well above V_{th}.

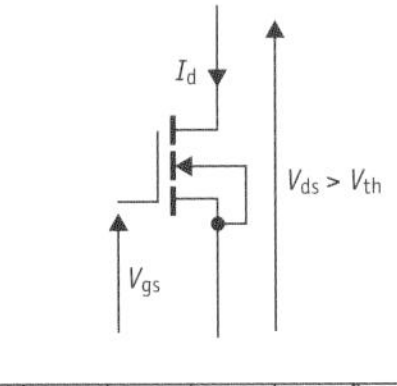

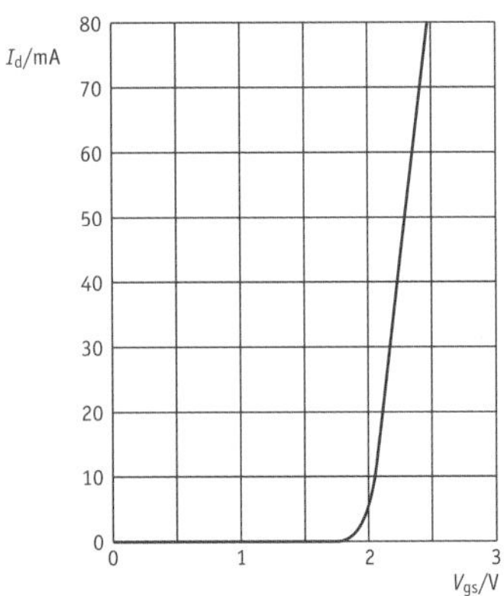

Fig 1.15 The $I_d - V_{gs}$ characteristic for a typical MOSFET where $V_{ds} > V_{th}$

MOSFET parameters

Once you know the values of the threshold voltage V_{th} and transconductance g_m of a MOSFET, you can work out how it will behave in the current sink region. They can be obtained from the $I_d - V_{gs}$ characteristic of Fig 1.15 as follows:

- Extrapolate the straight line back to the voltage axis and read off the value of V_{th}. In this case it is 2.0 V.
- Read off the drain current for any given gate–source voltage. For example, $I_d = 80$ mA when $V_{gs} = 2.5$ V.
- Use these values to determine the transconductance g_m.

$g_m = ?$

$V_{th} = 2.0$ V

$V_{gs} = 2.5$ V

$I_d = 80 \text{ mA} = 80 \times 10^{-3}$ A

$$I_d = g_m(V_{gs} - V_{th})$$

$$g_m = \frac{I_d}{V_{gs} - V_{th}} = \frac{80 \times 10^{-3}}{2.5 - 2.0} = 0.16 \text{ S or } 160 \text{ mS}$$

Constant current

Here's an example of the parameters applied to a useful MOSFET circuit. Take a look at Fig 1.16.

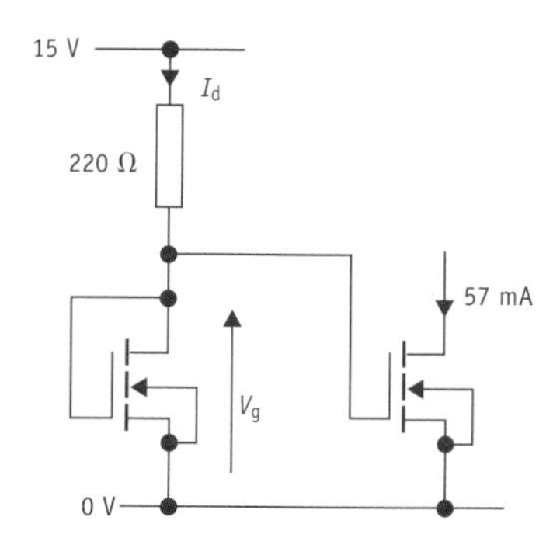

Fig 1.16 A pair of MOSFETs arranged as a constant current sink

The circuit is not complete – the component connected to the drain of the right-hand MOSFET is not shown. This is because it does not matter what it is. Provided that the drain is above 2 V, the drain current is completely fixed by the two MOSFET parameters, the 220 Ω drain resistor and the voltage of the supply rail. Here is the procedure for calculating the value of that drain current when $V_{th} = 2.0$ V and $g_m = 0.16$ S:

- Use $V = IR$ for the resistor: $15 - V_g = I_d \times 220$.
- Rearrange to find the voltage at the gate: $V_g = 15 - 220I_d$
- The source is held at 0 V, so $V_{gs} = V_g = 15 - 220I_d$
- Insert this into the MOSFET model $I_d = g_m(V_{gs} - V_{th})$: $I_d = 0.16 \times (15 - 220I_d - 2.0)$
- Multiply out the numbers: $I_d = 0.16 \times (13 - 220I_d) = 2.08 - 35.2I_d$
- Rearrange: $I_d + 35.2I_d = 2.08$, so $I_d = \dfrac{2.08}{1 + 35.2} = 5.7 \times 10^{-2}$A or 57 mA

Both MOSFETs have the same value of V_{gs} so they will have the same value of I_d. So the drain current of the right-hand MOSFET is a rock-steady 57 mA, provided that the drain is above 2 V. This constant-current sink is a key component in allowing the MOSFET to become a perfect 'voltage follower'.

Voltage follower

You will recall that drawing current from a signal source reduces its amplitude. This is because every signal source has an output impedance. Any current in that resistance results in a voltage drop, reducing the amplitude of the signal voltage reaching the next stage. The output impedance of many useful signal sources (such as tuned circuits in radios and nerve signals in humans) have very large output impedances, so there is a big demand for a voltage follower which can copy a signal without drawing any current from its source. Fig 1.17 shows a simple MOSFET voltage follower which does just that.

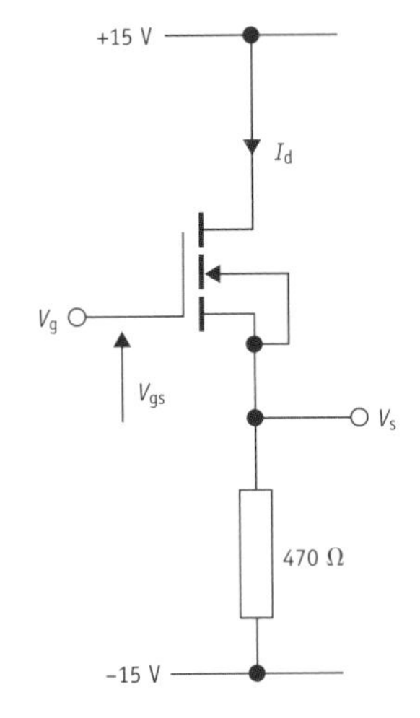

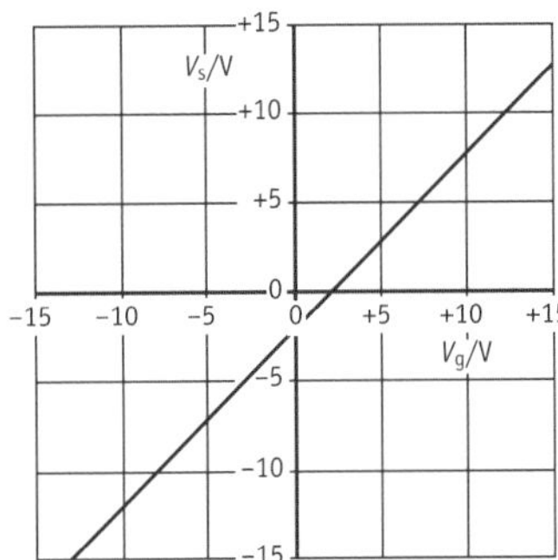

Fig 1.17 A simple MOSFET voltage follower

So how does it work? Consider what happens to the voltage at the source V_s as the voltage at the gate V_g is swept from −15 V to +15 V. The drain current will remain zero until V_{gs} reaches the threshold voltage of 2.0 V. So V_s remains at −15 V until V_g reaches −13 V. Once V_g is above −13 V, V_s will follow it, but with an **offset** of $-V_{gs}$. This offset will start off at the value of the threshold voltage, 2.0 V and will increase as the drain current increases. The value of V_s for $V_g = +15$ V can be worked out as follows.

- Use $V = IR$ for the resistor: $V_s - (-15) = I_d \times 470$
- Rearrange to find the voltage at the source: $V_s = 470I_d - 15$
- The gate is held at 15 V, so $V_{gs} = 15 - V_s$
- Insert the formula for V_s: $V_{gs} = 15 - (470I_d - 15) = 15 - 470I_d + 15 = 30 - 470I_d$
- Insert this into the MOSFET model $I_d = g_m(V_{gs} - V_{th})$:
  $I_d = 0.16 \times (30 - 470I_d - 2.0)$
- Multiply out the numbers: $I_d = 0.16 \times (28 - 470I_d) = 4.48 - 75.2I_d$
- Rearrange: $I_d + 75.2I_d = 4.48$, so $I_d = \dfrac{4.48}{1 + 75.2} = 5.9 \times 10^{-2}$A
- Calculate the source voltage:
  $V_s = 470I_d - 15 = 470 \times 5.9\times10^{-2} - 15 = 12.6$ V

So the value of V_{gs} creeps up from 2.0 V to 2.4 V as V_g sweeps from −13 V to +15 V. This means that the circuit has a voltage gain of slightly less than one, with an offset of about −2.2 V in the middle of its range. However, there is no current in the gate, so all of the signal is transferred from the source. Furthermore, the drain current is 27 mA in the middle of the range, so the current in the output terminal can be at least 1 mA before it has much effect on the output signal.

AC signals

The offset can be easily removed if the follower only has to copy a.c. signals, as shown in Fig 1.18.

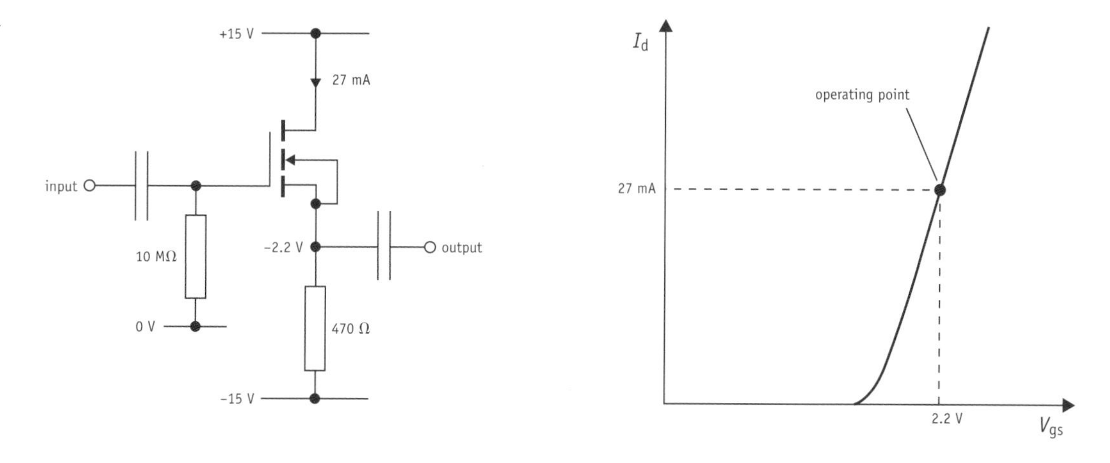

Fig 1.18 The 10 MΩ resistor holds the MOSFET at a suitable operating point

The circuit is **biased** in the linear part of its characteristic by using a 10 MΩ resistor to hold the gate at 0 V. You can verify for yourself that this leaves V_{gs} at 2.2 V and I_d at 27 mA. A coupling capacitor allows a.c. signals at the input terminal into the gate, but blocks any d.c. component of the input signal. As the a.c. signal varies the voltage of the gate above and below 0 V, the **operating point** moves up and down the characteristic, resulting in a change of drain current. This change of current in the 470 Ω source resistor results in an a.c. signal at the souce which passes through another coupling capacitor to the output terminal. The −2.2 V d.c. signal at the source is blocked by the capacitor, so it doesn't get to the output terminal.

Shortcomings

The presence of a bias resistor at the gate reduces the input impedance of the circuit from almost infinity to only 10 MΩ. This means that it is not much good for signal sources with an output impedance of more than 10 MΩ. More seriously, it is only distortion-free if the MOSFET can be accurately modelled by the formula $I_d = g_m(V_{gs} - V_{th})$. This is only true for real MOSFETs when the drain current is large. This can be wasteful, as it results in a lot of heating. For example, consider the heating power of the MOSFET when no a.c. signals are present.

$P = ?$

$V = 15 - (-2.2) = 17.2\text{ V}$

$I = 27\text{ mA} = 27 \times 10^{-3}\text{ A}$

$P = IV = 27 \times 10^{-3} \times 17.2 = 0.46\text{ W}$

This is less than the power rating of a typical small MOSFET (800 mW), but it is still a problem for circuits which run off batteries. Large currents shorten battery lifetimes, so they should be avoided if possible.

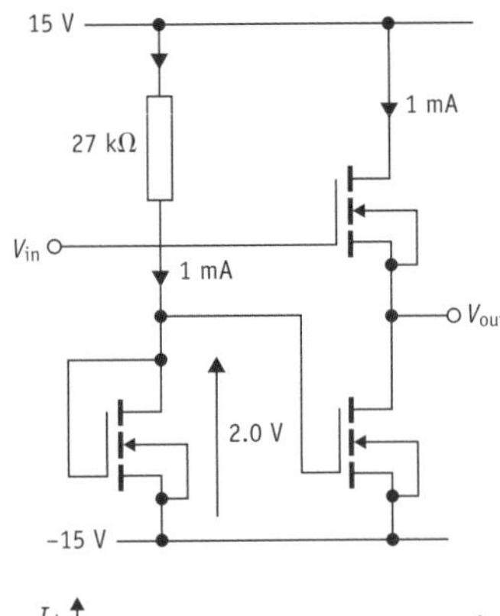

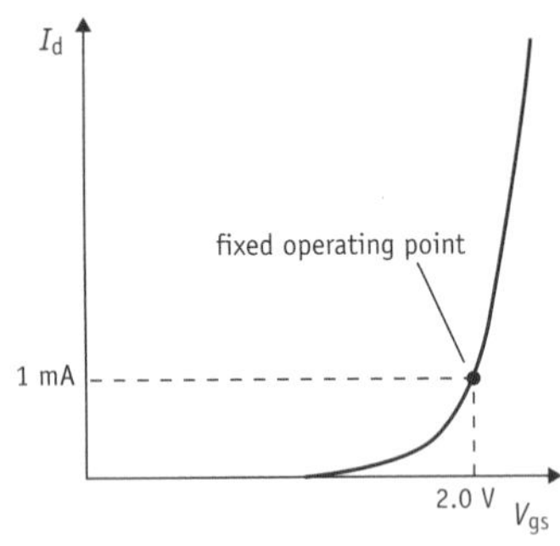

Fig 1.19 A constant offset voltage follower

Constant offset

Replacing the source resistor of a voltage follower with a MOSFET configured as a constant current sink allows the construction of a linear follower which doesn't require a large current, as shown in Fig 1.19.

The two MOSFETs connected to the bottom supply rail are arranged as a constant current sink for the MOSFET that is connected to the upper rail. The gate of the bottom left-hand MOSFET will sit at about −13 V, with $V_{gs} \approx 2.0$ V, the threshold voltage. The current in the drain resistor will therefore be about 1 mA.

$I = ?$

$V = 15 - (-13) = 28$ V

$R = 27\ \text{k}\Omega = 27 \times 10^{+3}\ \Omega$

$$R = \frac{V}{I} = \frac{28}{27 \times 10^{+3}} = 1.0 \times 10^{-3} \text{ A or 1 mA}$$

So the drain current of the third MOSFET will be a fixed 1 mA. This means that it has a fixed operating point on the $I_d - V_{gs}$ characteristic of Fig 1.19. As the gate moves up and down, the source will follow it with a fixed offset of 2.0 V corresponding to the fixed drain current of 1 mA. The result is a perfectly linear follower, regardless of the fact that it is operating on a very non-linear part of the MOSFET characteristic. Provided that V_{out} stays above −13 V, $V_{out} = V_{in} - 2.0$ V.

Matched MOSFETs

For the circuit to work, all three MOSFETs must be identical. This can only be done by etching the MOSFETs from the same mask on the same crystal of silicon – in other words, this is the type of circuit which is only possible on an integrated circuit.

No offset

The addition of another pair of MOSFETs and an op-amp circuit results in a voltage follower which has near-perfect behaviour. The circuit shown in Fig 1.20 has the following properties:

- It draws virtually no current from the source of the signal V_{in}.
- The output current can be several milliamps.
- $V_{out} = V_{in}$ over the large range of −13 V to +13 V.
- It can handle both d.c. and a.c. signals.

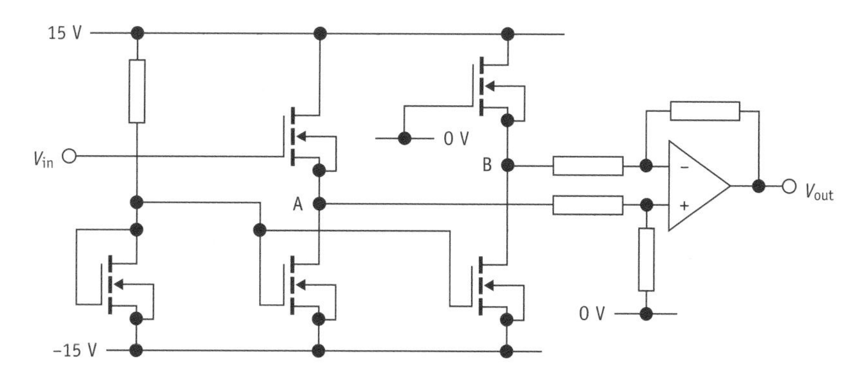

Fig 1.20 A linear follower with no offset for signals between ±13 V

So how is this perfection accomplished? Consider the signals at A and B.

$$V_A = V_{in} - V_{gs} \qquad V_B = 0 - V_{gs}$$

The output of the op-amp circuit is the difference of the signals at its two inputs.

$$V_{out} = V_A - V_B = (V_{in} - V_{gs}) - (0 - V_{gs}) = V_{in} - V_{gs} + V_{gs} = V_{in}$$

Of course, this circuit is not practicable with discrete components. All five MOSFETs must be etched onto the same slice of silicon so that they have identical behaviour. The op-amp with its four identical resistors might as well be etched onto the silicon as well, resulting in the five-terminal device shown in Fig 1.21.

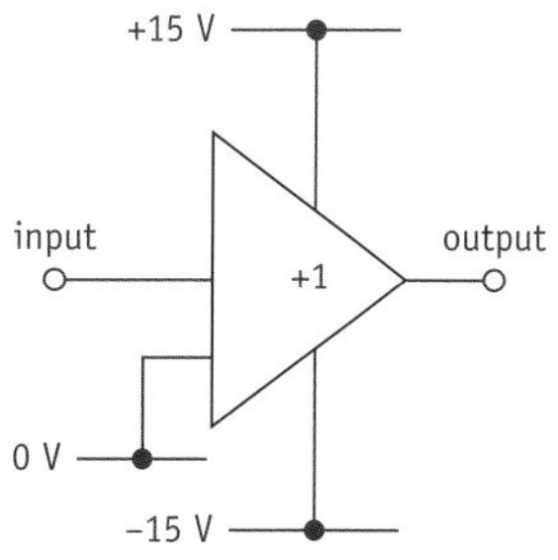

+15 V
input
+1
output
0 V
−15 V

Fig 1.21 The circuit of Fig 1.20 works best as a single integrated circuit

Difference amplifier

The signal at the output V_{out} of a difference amplifier is the difference between the signals at its inputs, V_A and V_B.

$$V_{out} = V_A - V_B$$

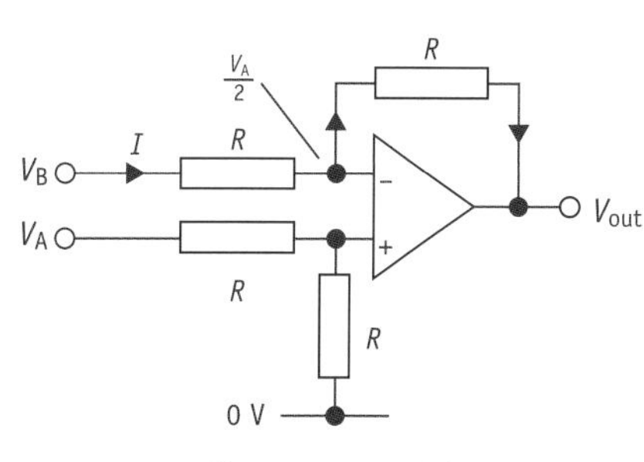

Fig 1.22 Difference amplifier

If you study the circuit in Fig 1.22, you will notice that it employs negative feedback, with a resistor connecting the output to the inverting input. This means, of course, that the value of V_{out} always settles to a value which leaves the two op-amp inputs at the same voltage. This idea leads to the difference amplifier formula above, as follows.

- The resistors at the non-inverting input hold it halfway between V_A and 0 V: $V_+ = \frac{V_A}{2}$
- Both op-amp inputs have the same voltage: $V_- = V_+$ so $V_- = \frac{V_A}{2}$
- Apply $V = IR$ to the resistor between V_- and V_B: $V_B - \frac{V_A}{2} = IR$
- Do the same for the feedback resistor: $\frac{V_A}{2} - V_{out} = IR$
- Equate the two expressions for IR: $V_B - \frac{V_A}{2} = \frac{V_A}{2} - V_{out}$
- Rearrange to find V_{out}: $V_{out} = \frac{V_A}{2} + \frac{V_A}{2} - V_B = V_A - V_B$

Of course, this only works when the op-amp output is not saturated.

MOSFET inverter

MOSFETs can do more than just make a copy of an a.c. signal. They can also increase its amplitude. The trick is to insert a resistor between the drain and the top supply rail, as shown in Fig 1.23.

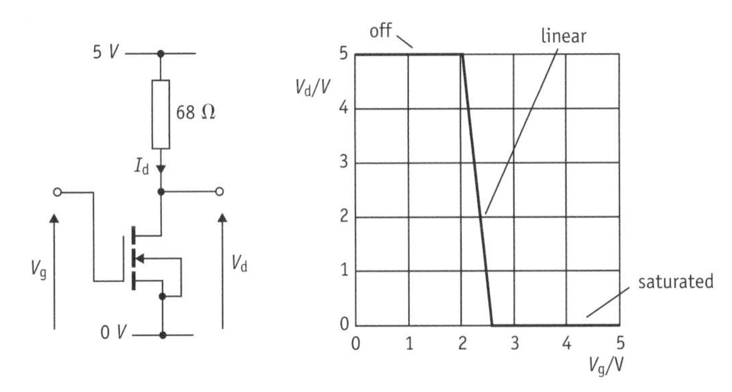

Fig 1.23 Characteristic of a MOSFET inverter

Configuring a MOSFET in this way makes it behave like a NOT gate, or inverter. There are three distinct regions of the characteristic.

- The MOSFET is **off** when V_g is below the threshold voltage (2.0 V in this case), so the drain is pulled up to 5 V by the resistor.
- Once V_g is above the threshold voltage, the MOSFET turns on and becomes **linear**. Increasing the gate voltage increases the drain current. This, in turn, increases the voltage drop across the resistor, lowering the voltage across the drain.
- The MOSFET **saturates** when the drain current is large enough to give a 5 V drop across the resistor. Any further increase in V_g has no effect on the value of V_d.

Amplifiers exploit the linear region of the inverter characteristic. The sharper the fall of the drain voltage, the larger the gain.

Gate bias

Provided that it is biased in the linear region of the characteristic, the MOSFET can amplify a.c. signals. This requires two additions to the inverter circuit:

- A voltage divider to hold the gate somewhere between 2.0 V and 2.5 V.
- Coupling capacitors to allow a.c. signals in and out of the circuit.

These are shown in Fig 1.24.

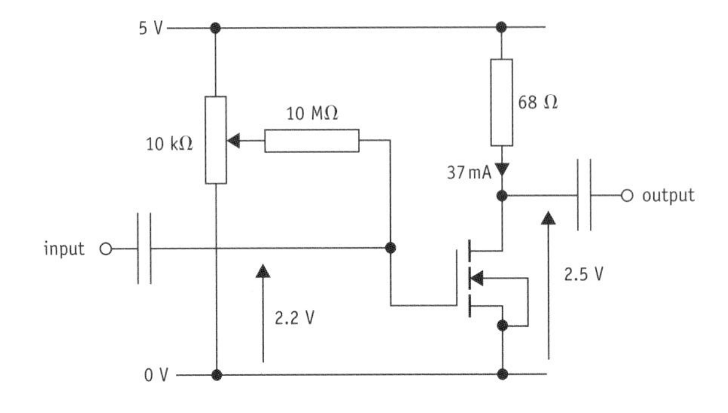

Fig 1.24 MOSFET amplifier

Notice that there is a 10 MΩ resistor between the gate and the potentiometer. This allows a.c. signals passing through the left-hand capacitor to see a large input impedance. Since there is no current in the gate, there will be no voltage drop across the 10 MΩ resistor in the absence of a.c. signals, so the wiper and gate will be at the same voltage. The gate voltage needs to be just right, so that the drain sits halfway between the two supply rails. This operating point allows the amplifier to produce distortion-free signals with amplitudes of up to 2.5 V, taking full advantage of the 5 V supply.

Bias calculation

The bias voltage required at the wiper is calculated in two steps.

- Calculate the drain current when V_d is at +2.5 V, halfway between 5 V and 0 V.

$I_d = ?$

$R = 68\ \Omega$

$V = 5 - 2.5 = 2.5$ V

$$I = \frac{V}{R} = \frac{2.5}{68} = 3.7 \times 10^{-2}\text{A or } 37\text{ mA}$$

- Use the two MOSFET parameters to calculate the gate voltage required.

$V_{gs} = ?$

$I_d = 37$ mA $= 37 \times 10^{-3}$ A

$g_m = 0.16$ S

$V_{th} = 2.0$ V

$$I_d = g_m(V_{gs} - V_{th})$$

$$V_{gs} - V_{th} = \frac{I_d}{g_m} = \frac{37 \times 10^{-3}}{0.16} = 0.23$$

$$V_{gs} = 0.23 + 2.0 = 2.23\text{ V}$$

You could use a pair of resistors as a fixed voltage divider to generate the 2.2 V required, but in practice it is easier to use a potentiometer to set the drain at 2.5 V by trial-and-error.

Signals

The graphs of Fig 1.25 show typical signals at the gate and drain of the amplifier when it has been correctly biased.

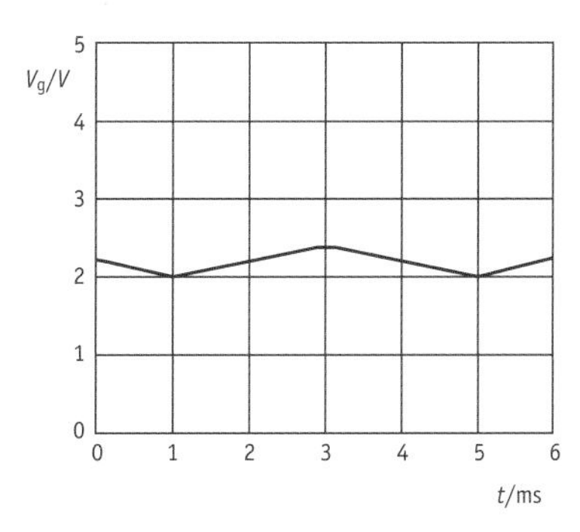

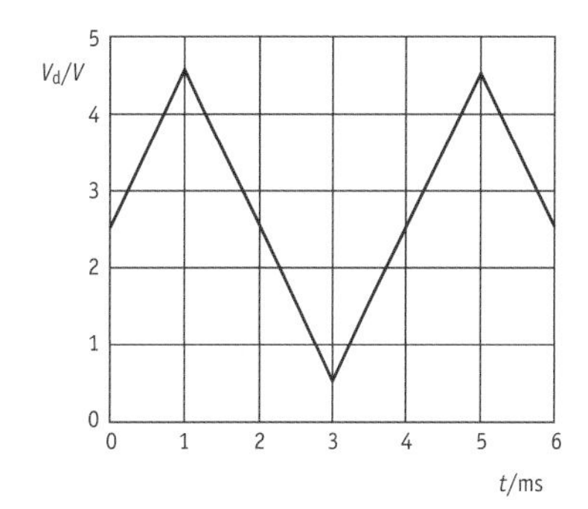

Fig 1.25 Signals at the gate and drain with correct bias

The a.c. signal at the gate is a triangle wave with an amplitude of only 0.2 V, centred on the d.c. signal of 2.23 V from the potentiometer. The corresponding signal at the drain has an a.c. component of amplitude 2.0 V, centred on 2.5 V. It has the same shape, but is inverted – a rise in voltage at the gate results in a much larger fall at the drain. The overall signal gain is about −10.

Gain

The gain G of an a.c. amplifier can be calculated from this formula.

$$G = \frac{\Delta V_{out}}{\Delta V_{in}}$$

ΔV_{out} is the instantaneous voltage of the a.c. signal at the output and ΔV_{in} is the corresponding voltage of the signal at the input. The graph of Fig 1.26 can be used to calculate the gain of an amplifier from its $V_d - V_g$ characteristic.

Fig 1.26 A small rise of gate voltage results in a much larger drop of drain voltage

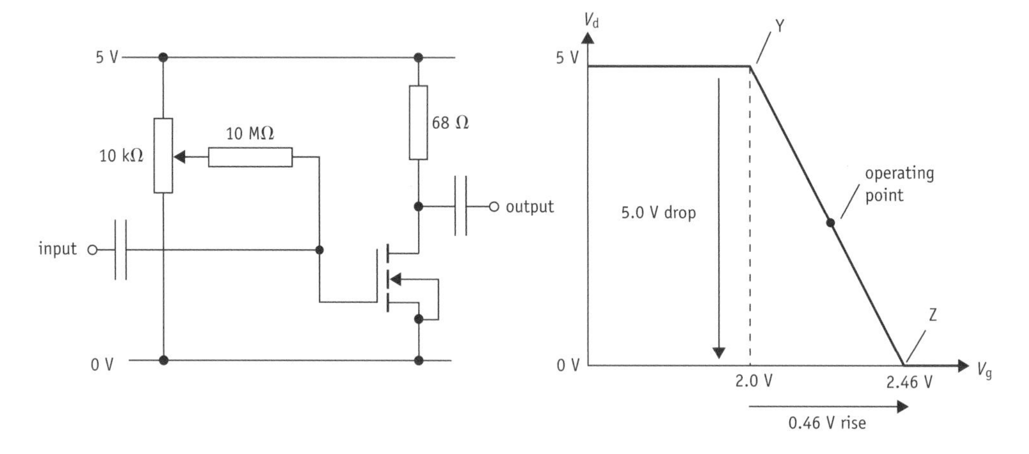

In the absence of a.c. signals at the amplifier input, the potentiometer holds the gate at +2.23 V, halfway down the linear region. Now suppose that an a.c. signal of amplitude 0.23 V is added to this d.c. signal. This makes the amplifier oscillate between Y and Z at opposite ends of the linear region. As the gate rises from 2.00 V to 2.46 V the drain falls from 5.0 V to 0.0 V. The gain of the amplifier can be found by considering the voltages of the signals at input and output when the system is momentarily at Z.

$G = ?$

$\Delta V_{in} = 2.46 - 2.23 = +0.23$ V

$\Delta V_{out} = 0.0 - 2.5 = -2.5$ V

$$G = \frac{\Delta V_{out}}{\Delta V_{in}} = \frac{-2.5}{+0.23} = -11$$

The gain G of a MOSFET amplifier depends only on its transconductance g_m and the resistor R_d between the drain and the top supply rail. Here is the formula.

$G = ?$

$g_m = 0.16$ S

$R_d = 68\ \Omega$

$$G = g_m R_d$$

$$G = -g_m R_d = -0.16 \times 68 = -11$$

You will find out where this formula comes from on the next page.

Large currents

So what are the limits on the gain that you can get from a MOSFET amplifier? At first glance, the gain can be as large as you like. Increasing the value of R_d should increase the value of G. Unfortunately, this reduces the value of the drain current. The actual gain of a MOSFET amplifier will only agree with the calculated value if its linear region is a straight line. In other words, if it obeys the equation $I_d = g_m(V_g - V_{th})$. You will recall that this is only true if the drain current is large enough to keep V_{gs} well above V_{th}. In Fig 1.26, the drain current is 37 mA, so we are safe. Increasing the value of R_d from 68 Ω to 680 Ω to give a gain of −110, reduces the drain current to only 3.7 mA. So it is quite likely that, in practice, the gain will turn out to be a lot less than −110.

Gain formula

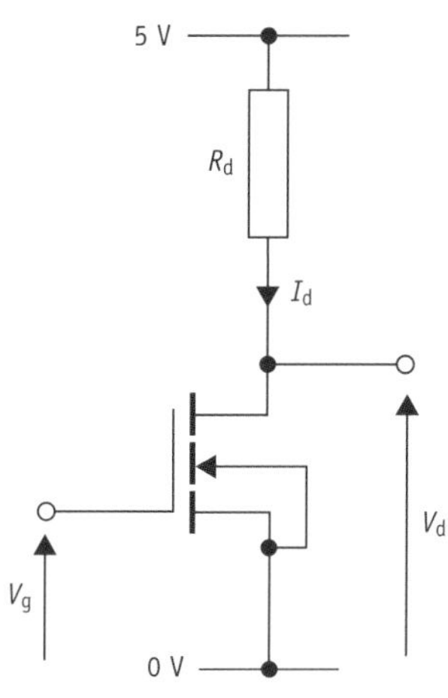

Fig 1.27 The heart of a MOSFET amplifier

Consider the circuit shown in Fig 1.27.

The gain formula for the MOSFET voltage follower can be obtained as follows:

- The drain current is fixed by the voltage at the gate: $I_d = g_m(V_g - V_{th})$
- Use $V = IR$ to find the voltage at the drain: $5 - V_d = I_d R_d$
- Eliminate I_d: $5 - V_d = g_m(V_g - V_{th})R_d = g_m R_d V_g - g_m R_d V_{th}$
- Rearrange to find V_d: $V_d = 5 + g_m R_d V_{th} - g_m R_d V_g$

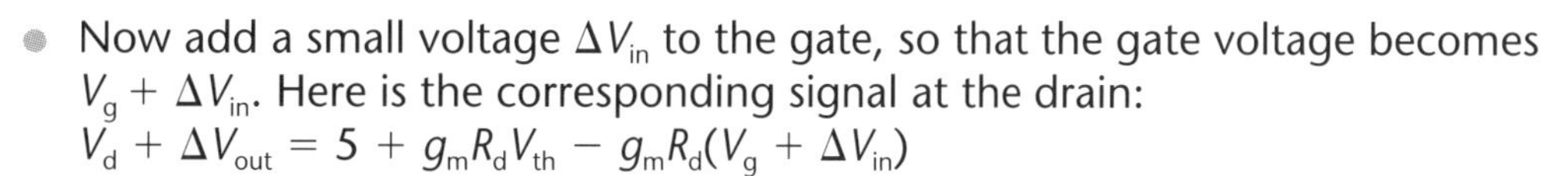

- Now add a small voltage ΔV_{in} to the gate, so that the gate voltage becomes $V_g + \Delta V_{in}$. Here is the corresponding signal at the drain: $V_d + \Delta V_{out} = 5 + g_m R_d V_{th} - g_m R_d(V_g + \Delta V_{in})$

- Multiply out the brackets: $V_d + \Delta V_{out} = 5 + g_m R_d V_{th} - g_m R_d V_g - g_m R_d \Delta V_{in}$
- Compare the two expressions for the drain voltage:

$$V_d = 5 + g_m R_d V_{th} - g_m R_d V_g$$

$$V_d + \Delta V_{out} = 5 + g_m R_d V_{th} - g_m R_d V_g - g_m R_d \Delta V_{in}$$

- Use the first expression to cancel down the second: $\Delta V_{out} = -g_m R_d \Delta V_{in}$
- Rearrange: $\dfrac{\Delta V_{out}}{\Delta V_{in}} = -g_m R_d$

Of course, this all assumes that the drain current is large enough for $I_d = g_m(V_g - V_{th})$ to be a good description of the MOSFET's behaviour.

Fixed bias

The MOSFET amplifier shown in Fig 1.28 uses a pair of fixed resistors to set the bias.

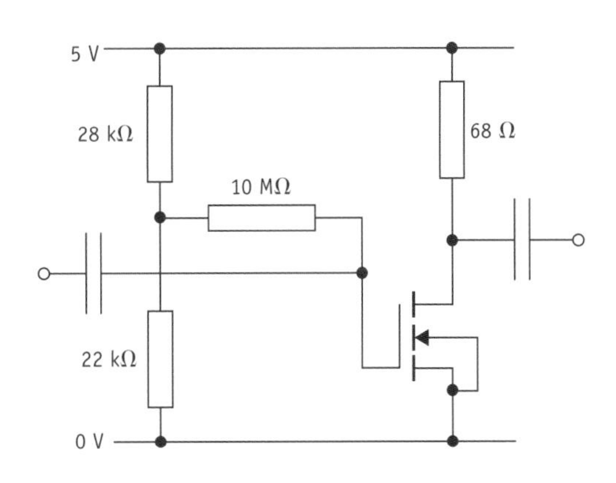

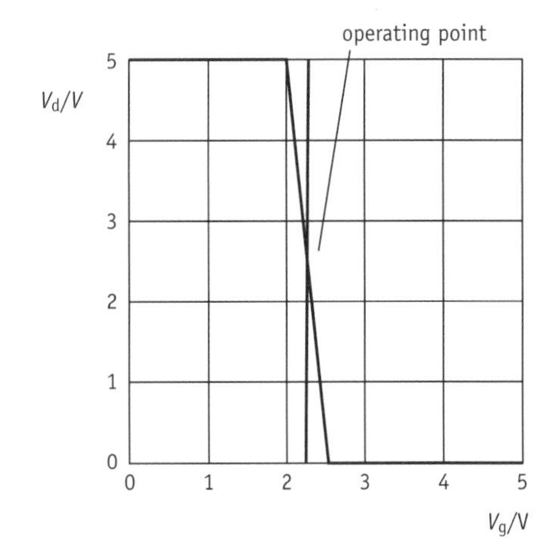

Fig 1.28 The operating point is where the two curves cross

You will recall that the gate voltage can be calculated in two steps. Start off by finding the current in the voltage divider.

$I = ?$

$V = 5\text{ V}$

$R = 22\text{ k}\Omega + 28\text{ k}\Omega = 50\times10^3\ \Omega$

$$I = \frac{V}{R} = \frac{5}{50 \times 10^3} = 1 \times 10^{-4}\text{A or } 100\ \mu\text{A}$$

Then find the voltage drop across just the bottom resistor.

$V = ?$

$I = 1 \times 10^{-4}\text{ A}$

$R = 22 \times 10^3\ \Omega$

$$V = IR = 1\times10^{-4} \times 22 \times 10^3 = 2.2\text{ V}$$

The vertical line on the graph of Fig 1.28 shows that the gate is held at this voltage, regardless of the voltage at the drain. The line passes through the centre of the linear region of the characteristic, so this is the operating point for the amplifier, with the gate voltage just a bit above the threshold voltage of 2.0 V.

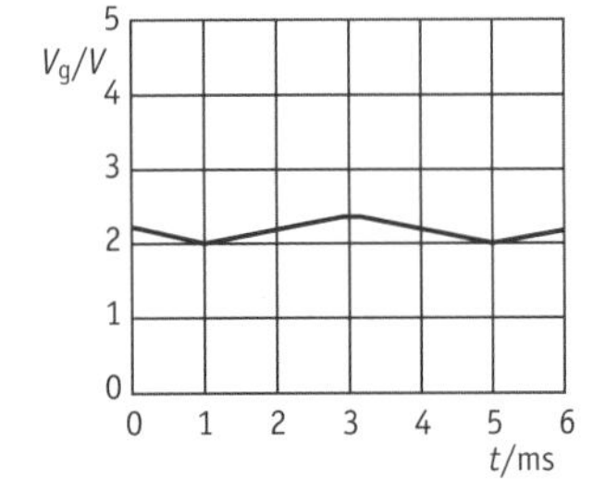

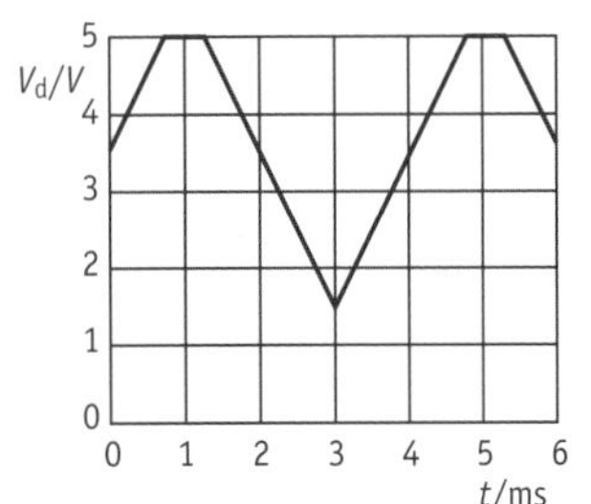

Fig 1.29 The output signal can saturate if the bias is incorrect

Clipping

Getting the bias correct is crucial. Suppose that the MOSFET in Fig 1.28 is replaced with one that has a different threshold voltage, 2.1 V instead of 2.0 V. The result is shown in the graphs of Fig 1.29. The signal at the gate is the same, but the signal at the drain has been shifted up by 1.0 V. The top of the output waveform has been **clipped** as the MOSFET enters the off region, distorting the signal. The solution, of course, is to raise the gate voltage by 0.1 V to 2.3 V, so that the two lines on the graph cross at $V_d = 2.5$ V once more. Since the values for g_m and V_{th} differ greatly from one MOSFET to another, the bias voltage at the gate will have to be different for each amplifier. That means either a different combination of fixed resistors, or the use of a potentiometer. Either way, getting the bias correct is a fiddly process.

Drain bias

There is a better way of providing the bias, one which will always work, whatever the values of the MOSFET threshold voltage and transconductance. The arrangement is shown in Fig 1.30.

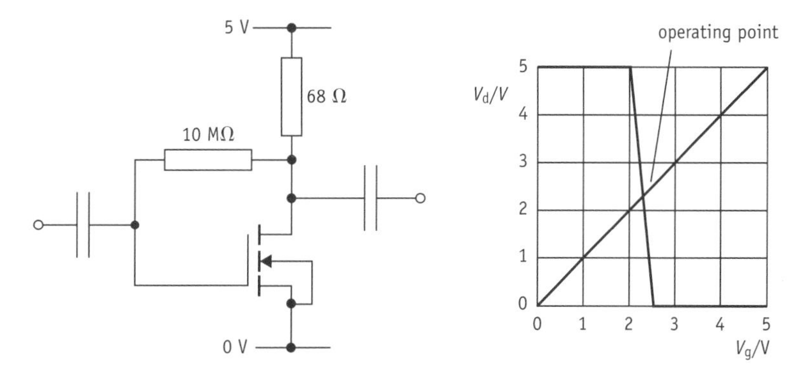

Fig 1.30 MOSFET amplifier with drain bias

The line at 45° on the graph represents the fact that the gate is connected to the drain by the 10 MΩ resistor. (As there is no current in the gate, there will be no voltage drop across that resistor, and $V_g = V_d$ in the absence of a.c. signals at the input.) That line must cross the linear region at some point, provided that the threshold voltage is less than 5 V. The operating point will be where the two lines cross. Of course, there is no guarantee that this will bias the drain halfway between the supply rails, but at least the operating point will be somewhere in the linear region, so the amplifier will always work without needing any adjustment by a human being! However, there is a price to be paid for this simplicity. The effective input impedance of the circuit is now only 1 MΩ, reduced from 10 MΩ because a.c. signals at the drain are ten times larger than a.c. signals at the gate.

Fixed gain

Although drain bias does guarantee that any MOSFET with a reasonable threshold voltage will be able to amplify a.c. signals without any need for adjustment, it doesn't alter the fact that each amplifier circuit will have a different gain. The circuit shown in Fig 1.31 gets around this as well.

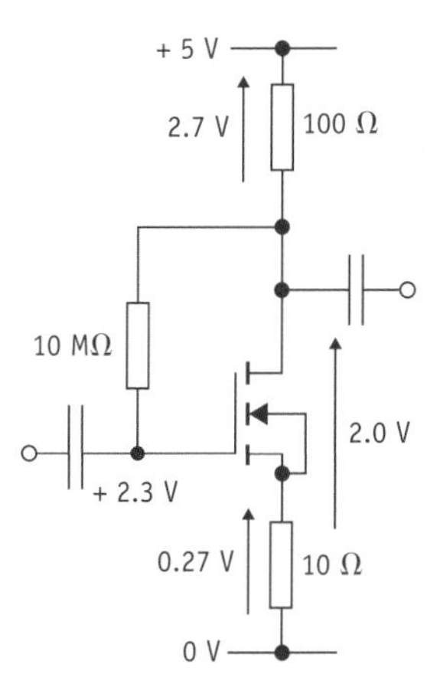

Fig 1.31 Amplifier with a gain of ten and automatically correct bias

The 10 MΩ bias resistor ensures that the drain–source voltage will be a bit above the threshold voltage of 2.0 V. This leaves about $5 - 2 = 3$ V to be shared by the 100 Ω and 10 Ω resistors, giving a drain current of about 27 mA. This places the MOSFET firmly on the linear region of its $I_d - V_{gs}$ characteristic (Fig 1.18). The voltage drop across the 100 Ω resistor is therefore about $27 \times 10^{-3} \times 100 = 2.7$ V, leaving the drain (and gate) at $5 - 2.7 = 2.3$ V.

So the bias works. What about the gain? Suppose that an a.c. signal at the input raises the gate voltage by 1 mV. The voltage at the source will also rise by 1 mV, as it's a voltage follower. However, the voltage at the drain will fall by 10 mV. This is because the voltage drop across the 100 Ω drain resistor is always ten times greater than the drop across 10 Ω source resistor. The 10:1 ratio of the resistors sets the gain to −10, regardless of the properties of the MOSFET. Unfortunately, with this circuit there isn't much room for the drain to move up and down. Output signals with amplitudes above 1 V are likely to be clipped, as the MOSFET is forced out of its current-sink region (Fig 1.14).

Questions

1.1 Variable resistors

1 This question is about a MOSFET which has a threshold voltage of 2.5 V.

(a) Draw the circuit symbol for the MOSFET, labelling the gate, drain and source.

(b) Describe how the drain–source resistance varies as the gate–source voltage is swept from 0 V to 5.0 V.

(c) 'For small enough values of the drain–source voltage, a MOSFET behaves like a voltage-controlled resistor.' Explain this statement, with the help of graphs of drain current against drain–source voltage for a variety of gate–source voltages.

2 The circuit shown in Fig 1.32 is a voltage-controlled volume control for audio signals.

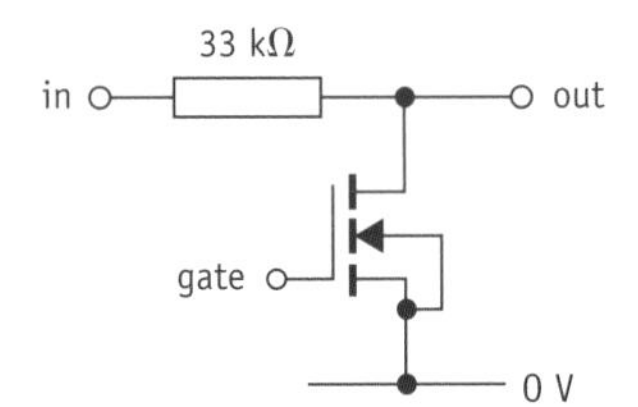

Fig 1.32 Circuit for question 2

(a) The MOSFET has a threshold voltage of 1.5 V. Sketch a graph to show how the drain–source resistance varies as the gate–source voltage is swept from 0 V to 5 V.

(b) Only a fraction of the audio signal at the input gets through to the output. Sketch a graph to show how the gain of the system varies as the gate–source voltage is swept from 0 V to 5 V. Explain its shape.

(c) Suggest the maximum amplitude audio signal at the input for which the system will operate correctly.

3 The circuit of Fig 1.33 is a voltage-controlled non-inverting amplifier.

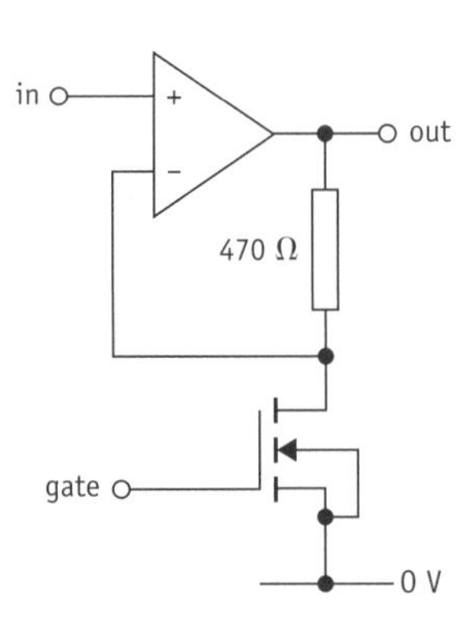

Fig 1.33 Circuit for question 3

(a) Show that the circuit has a gain of +16 when the drain–source resistance of the MOSFET is 32 Ω.

(b) The drain–source resistance R_{ds} of the MOSFET is fixed by its gate–source voltage V_{gs} by this formula.

$$R_{ds} = R_0 \frac{V_{th}}{V_{gs} - V_{th}}$$

The values of R_0 and V_{th} are 16 Ω and 2.0 V respectively. Use the formula to complete this table, where G is the gain of the circuit.

V_{gs} / V	R_{ds} / Ω	G
3.0		
4.0		
5.0		

(c) Show that the gain of the system is given by this approximate formula.

$$G = 15V_{gs} - 28.$$

4 The circuit in Fig 1.34 contains three analogue switches.

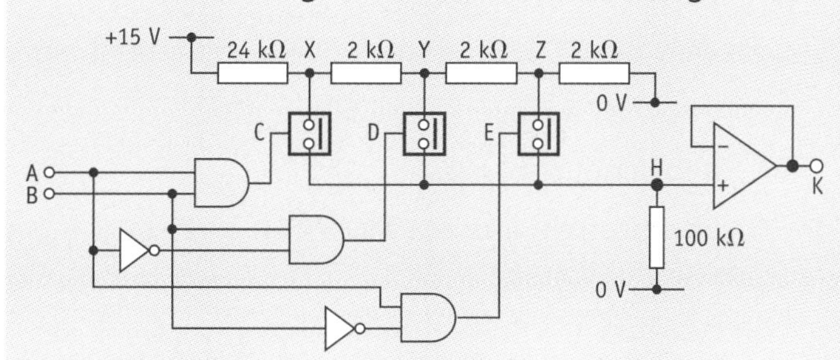

Fig 1.34 Circuit for question 4

(a) Describe the electrical properties of the analogue switch on the right.

(b) Explain why the voltage at K is the same as the voltage at H.

(c) By calculating the current in the resistor chain between +15 V and 0 V, determine the voltages at X, Y and Z.

(d) Complete this table for the system.

B	A	C	D	E	Signal at H
0	0	0	0	0	0.0 V
0	1				
1	0				
1	1				

(e) Suggest why the op-amp has been included in the circuit.

5 Analogue multiplexers and demultiplexers can control the flow of audio signals.

(a) Draw the circuit symbol for a two-input analogue multiplexer. Describe its behaviour.

(b) Draw a circuit to show how a two-input analogue multiplexer can be made from a pair of analogue switches and a NOT gate. Explain how the circuit operates.

(c) Explain why the circuit is both a multiplexer and a demultiplexer.

(d) Show how an eight-input analogue multiplexer can be made from seven two-input ones.

6 The circuit shown in Fig 1.35 has the behaviour of a tristate.

(a) Draw the circuit symbol for a tristate. With the help of a truth table, describe its electrical behaviour.

(b) Describe the differences in behaviour between an analogue switch and a tristate.

(c) Complete this table for the system shown in Fig 1.35.

E	A	X	Y	Q
1	1	1	0	1
				–
				–
				0

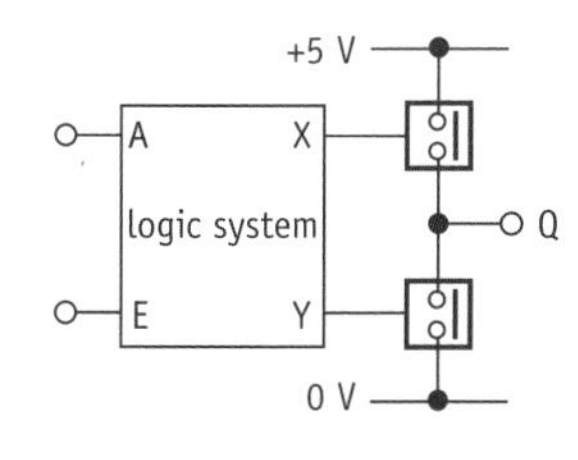

Fig 1.35 Circuit for question 6

(d) Design a circuit for the logic system using only NAND gates.

(e) Show how a tristate can be made from an OR gate and an analogue switch.

1.2 Amplifiers

Assume that the MOSFET in the following questions has these parameter values:

- threshold voltage $V_{th} = 1.6$ V
- transconductance $g_m = 0.12$ S

1 The drain current I_d of the MOSFET is given by $I_d = g_m(V_{gs} - V_{th})$ where V_{gs} is the gate–source voltage.

(a) Calculate the drain current when the gate–source voltage is 1.8 V.

(b) On the same set of axes, draw graphs to show how the drain current varies as the drain–source voltage is swept from 0 V to 6 V for gate–source voltages of 1.8 V and 2.0 V.

(c) Indicate the resistive and current sink regions of the graph. Describe the behaviour of the MOSFET in each of these regions.

(d) Draw a graph to show how the drain current varies as the gate–source voltage is swept from 0 V to 6V when the drain–source voltage is 2.0 V.

2 This question is about the voltage-follower circuit shown in Fig 1.36.

(a) By considering the voltage drop across the 220 Ω resistor, write down an expression linking the drain current I_d to the source voltage V_s.

(b) Use the MOSFET parameters V_{th} and g_m to write down another expression linking I_d and V_s.

(c) Use your answers from (a) and (b) to show that V_s is −1.9 V and I_d is 32 mA.

(d) Draw a graph to show how the source voltage varies as the gate voltage is swept from −9 V to +9 V.

(e) The circuit is a voltage follower with a gain of about 0.9, an offset of about −2 V and an input impedance of 22 MΩ. Justify this statement.

(f) Show the addition of two capacitors that would convert the system into a follower for a.c.signals. A sine wave signal of amplitude 0.5 V and frequency 500 Hz is fed into the system. On the same set of axes, sketch graphs to show the signals at the gate and source for two cycles of the signal.

(g) The MOSFET has a power rating of 500 mW. By calculating the power of the MOSFET when the gate is at 0 V, show that it will not be damaged.

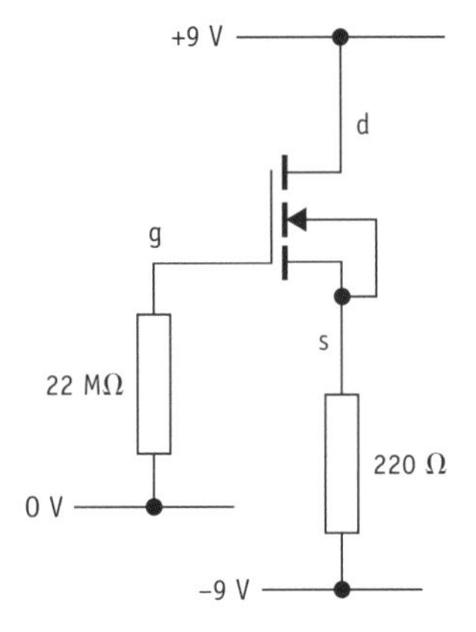

Fig 1.36 Circuit for question 2

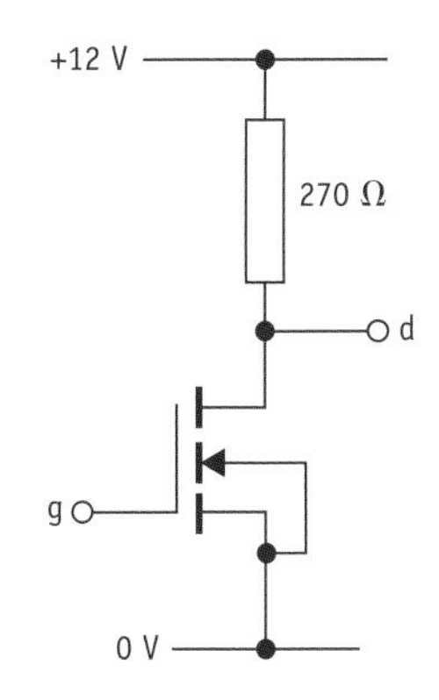

Fig 1.37: Circuit for question 3

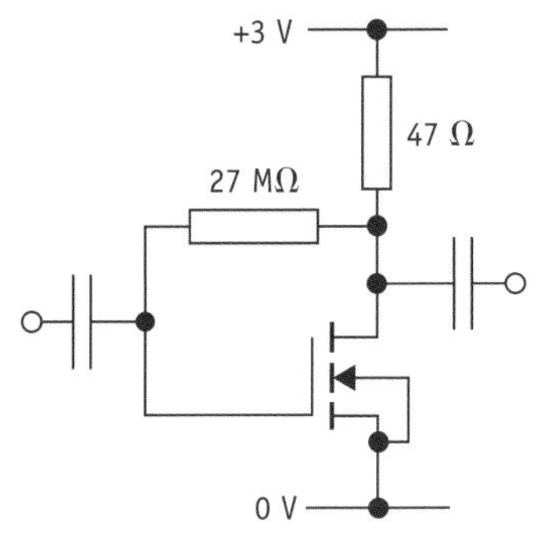

Fig 1.38: Circuit for question 4

3 This question is about the incomplete amplifier circuit shown in Fig 1.37.

(a) Show that the drain current I_d is 44 mA when the drain is saturated at 0 V.

(b) Use the formula $I_d = g_m(V_{gs} - V_{th})$ to calculate the smallest voltage at the gate which will saturate the MOSFET.

(c) Draw a graph to show how the voltage at the drain varies as the gate is swept from 0 V to +12 V.

(d) Explain why the gate should be held at +1.8 V to provide ideal bias for the amplifier.

(e) The amplifier is required to have an input impedance of 12 MΩ. Add three resistors and two capacitors to complete the circuit. Justify the values of all three resistors.

(f) Show that the amplifier has a gain of about −30.

(g) A sine wave signal of amplitude 0.1 V and frequency 2 kHz is fed into the amplifier. On the same set of axes, show the signals at the gate and drain for two cycles of the signal.

4 The amplifier shown in Fig 1.38 uses drain bias.

(a) Explain why, in the absence of a.c. signals at the input, the drain voltage is about halfway between the supply rails.

(b) Show that the gain of the amplifier is −6.

(c) Describe the advantages and disadvantages of drain biasing an amplifier.

Learning summary

By the end of this chapter you should be able to:

- sketch graphs to show how the drain current varies with drain–source voltage for different values of gate–source voltage
- use a MOSFET as a voltage-controlled resistor
- use analogue switches and multiplexers to control the flow of a.c. signals
- use tristates to control the flow of digital signals
- understand how to use a MOSFET as a voltage follower
- calculate drain current from the transconductance and threshold voltage
- understand how to use a MOSFET as an amplifier of a.c. signals

CHAPTER

2

Digital processing

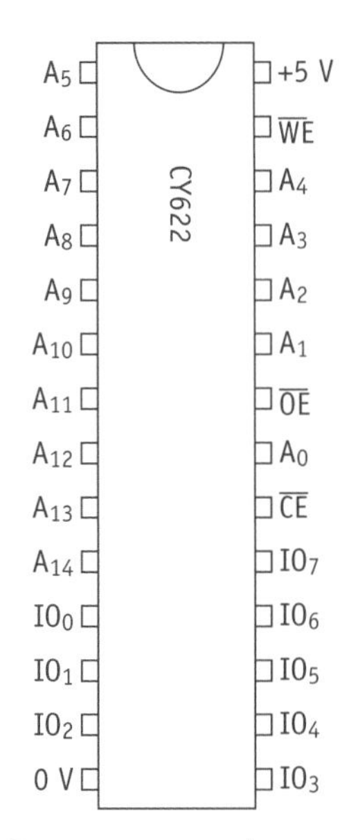

Fig 2.1 Pin connections for a 32K×8-bit memory i.c

2.1 Memory modules

You suddenly need to contact someone. What do you do? The chances are that you will reach for your mobile phone. Tap the first few initials of their name into the keypad and up comes their number on the screen. Squeeze the blue button with the phone symbol and, if their phone is on, the wonders of digital wireless technology allow you to communicate with them almost immediately. Of course, several systems have to work together to make this possible, but the first step (finding the number to call) relies on your phone having a **memory**.

Inputs, outputs

Any useful memory is a very large logic system, with many inputs and outputs. The pinout of a typical memory chip is shown in Fig 2.1.

The 28 connections to the chip fall into four distinct classes.

- Two connections link to the supply rails at +5 V and 0 V.
- The 15 inputs from the **address bus** are labelled A_{14} to A_0. These take in a binary word which specifies the **location** in the memory of the word to be accessed.
- The 8 connections to the **data bus** are labelled IO_7 to IO_0. They allow the two-way transfer of bytes in and out of each memory location.
- Three inputs from the **control bus** are labelled $\overline{CE}$, $\overline{OE}$ and $\overline{WE}$. Between them, these control the behaviour of the memory. For example, pulling only $\overline{CE}$ and $\overline{OE}$ makes the memory output a byte onto the data bus. Notice that the bar over the symbol tells you that each of the control bus signals is **active-low**. Their default setting is high, and have to be pulled low to accomplish their function.

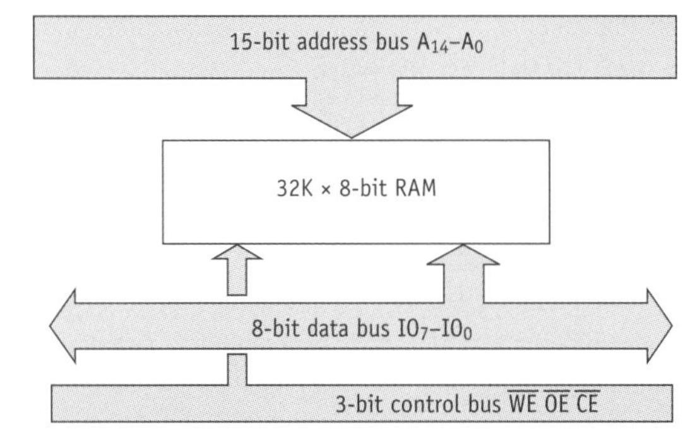

Fig 2.2 Each memory module is connected to three different buses

You have probably realised that the term **bus** refers to a set of parallel wires on a circuit board, leading to and from a memory module. Two of the buses are **unidirectional**, carrying information in one direction only. Only the data bus is **bidirectional**, carrying information towards or away from the module, according to the state of the control bus. The different types of bus are evident in the block diagram of Fig 2.2. The term RAM stands for **random access memory**, meaning that locations can be read from, or written to, in any order.

Reading and writing

Suppose that you wanted to **read** a word from the memory. Here is the sequence of operations required.

- Place the address of the location on the address bus.
- Wait for the location to be accessed by the **address decoder** in the memory. This may take up to 100 ns, depending on the number of different locations in the memory.
- Enable the memory by pulling the $\overline{\text{CE}}$ (**not-chip-enable**) pin low. This activates the chip, preparing it to access the data bus.
- Pull the $\overline{\text{OE}}$ (**not-output-enable**) line low. This enables the tristates on the eight data lines of the memory, allowing the word stored in the accessed location to control the state of the eight lines of the data bus.

It is then up to some other device connected to the data bus to swallow the byte while $\overline{\text{OE}}$ is low. Once $\overline{\text{OE}}$ goes high again, the memory is disconnected from the data bus as each tristate at the data outputs reverts to an open circuit state. The whole process is summarised in the timing diagram of Fig 2.3.

Fig 2.3 Reading from a memory

Note that the timing diagram shows the state of buses as well as lines. As usual, time runs from left to right, with up as high and down as low for the lines. A different convention is used for the buses. A pair of horizontal lines represents a stable word, lines at 45° represent changes in the word and a single horizontal line represents an open circuit state.

The operation of storing a word in memory is called **writing**. This is how you do it.

- Place the location address on the address bus.
- Place the word to be stored on the data bus.
- Pull $\overline{\text{CE}}$ low to activate the chip.
- Pulse $\overline{\text{WE}}$ (**not-write-enable**) low. The word on the data bus is stored in the memory during the rising edge at the end of the pulse.

The process is summarised in the timing diagram of Fig 2.4.

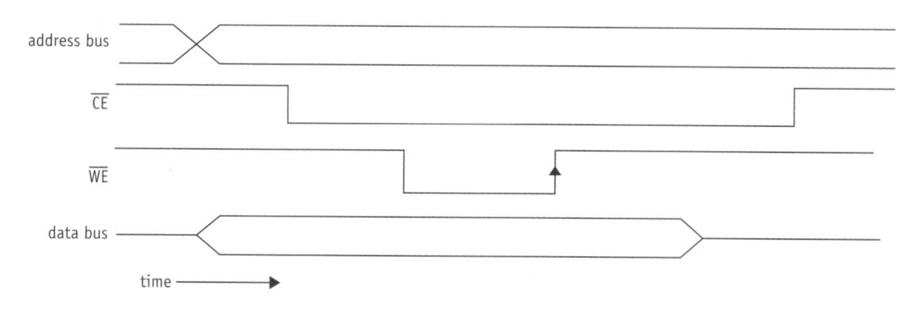

Fig 2.4 Writing to a memory

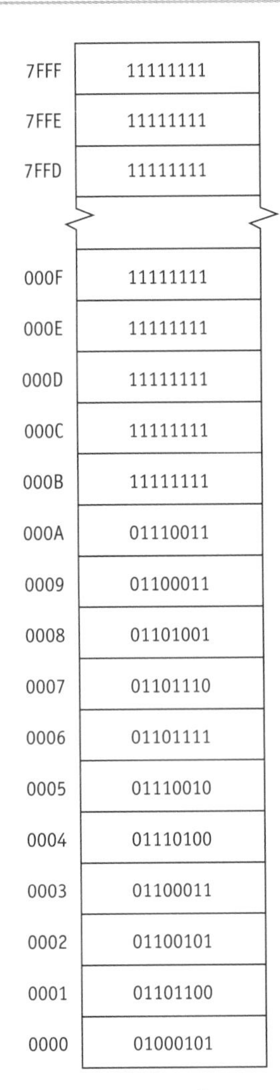

Fig 2.5 Memory map for a 32K×8-bit memory

Memory map

Fig 2.5 uses a **memory map** to represent the internal organisation of a memory.

Each box represents a single memory location, with the boxes stacked one on top the other. Each location stores a word, shown in binary inside the box. The locations near the bottom of the memory contain information (the word 'Electronics' coded in ASCII), but each location further up holds 11111111, the default setting before it has been written to. Note that even when a location doesn't store anything useful it still has to contain an 8-bit word – the default setting for each bit is 1 or 0 depending on the technology used to build the memory. Since the data bus is eight bits wide, each location holds an 8-bit word (a **byte**).

To the left of each box is shown its address, in hexadecimal. The address bus is 15 bits wide, so it makes sense to group the bits in groups of four (a **nibble**) and code each group from 0 to F. So the address of the topmost location in the memory has an address of 0111 1111 1111 1111 or 7FFF. The number of different locations in the memory M is fixed by the number of address lines N with this formula.

$$M = 2^N$$

In this case $N = 15$, so $M = 2^{15} = 32\ 768$. This is often shortened to 32K where $1\ \text{K} = 2^{10} = 1024$. (Larger memories are often quoted in units of M, where $1\ \text{M} = 2^{20} = 1\ 048\ 576$.)

Bigger words

Computers routinely require memories whose size vastly exceeds what can be crammed into a single integrated circuit. It is therefore often necessary to combine several smaller memories to make one larger memory. For example, suppose that you need to assemble a 64K×12-bit memory from just 16K×4-bit memory modules. How can it be done? The first stage is to assemble three modules to make a memory which can read or write 12-bit words, as shown in Fig 2.6. Each module is connected to the same control and address bus, but the data bus is divided up between them. The whole package behaves like a single 16K×12-bit memory.

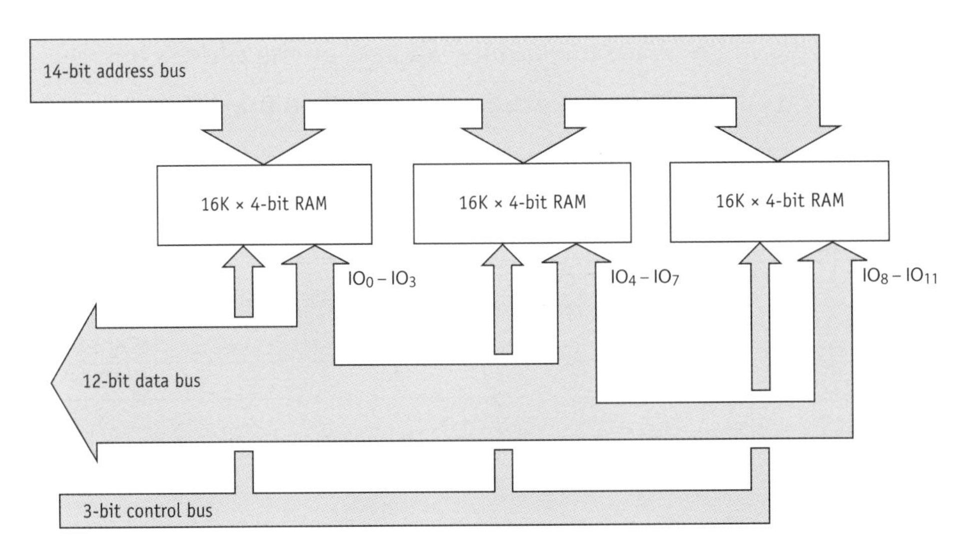

Fig 2.6 Three modules combined to make one 16K×12-bit memory

More addresses

The next stage in assembling the 64K×12-bit memory is to decide how many address lines it will need. By trial-and-error, $2^{16} = 65\,536$, so it will need a 16-bit address bus, with lines labelled from A_{15} to A_0. Fig 2.7 shows how this can be implemented with four of the 16K×12-bit memories of Fig 2.6.

The memory map of Fig 2.8 shows how the 64K locations are shared amongst the four RAMs. Each RAM is accessed by a different range of addresses to all of the others, thanks to the demultiplexer. The truth table shows how the demultiplexer directs the $\overline{CE}$ signal from the control bus to just one of the 16K×12-bit RAMs, depending on the state of the top two bits of the address bus, A_{15} and A_{14}. The pull-up resistors ensure that the $\overline{CE}$ signal for each RAM is high when it is not selected by the demultiplexer.

A_{15}	A_{14}	$\overline{CE_3}$	$\overline{CE_2}$	$\overline{CE_1}$	$\overline{CE_0}$
0	0	1	1	1	$\overline{CE}$
0	1	1	1	$\overline{CE}$	1
1	0	1	$\overline{CE}$	1	1
1	1	$\overline{CE}$	1	1	1

Note that the bus structure for the new large RAM is similar to that of the original smaller 16K×4-bit RAMs that it is made from. This means that RAM can be scaled up to any size that you like for very little design effort.

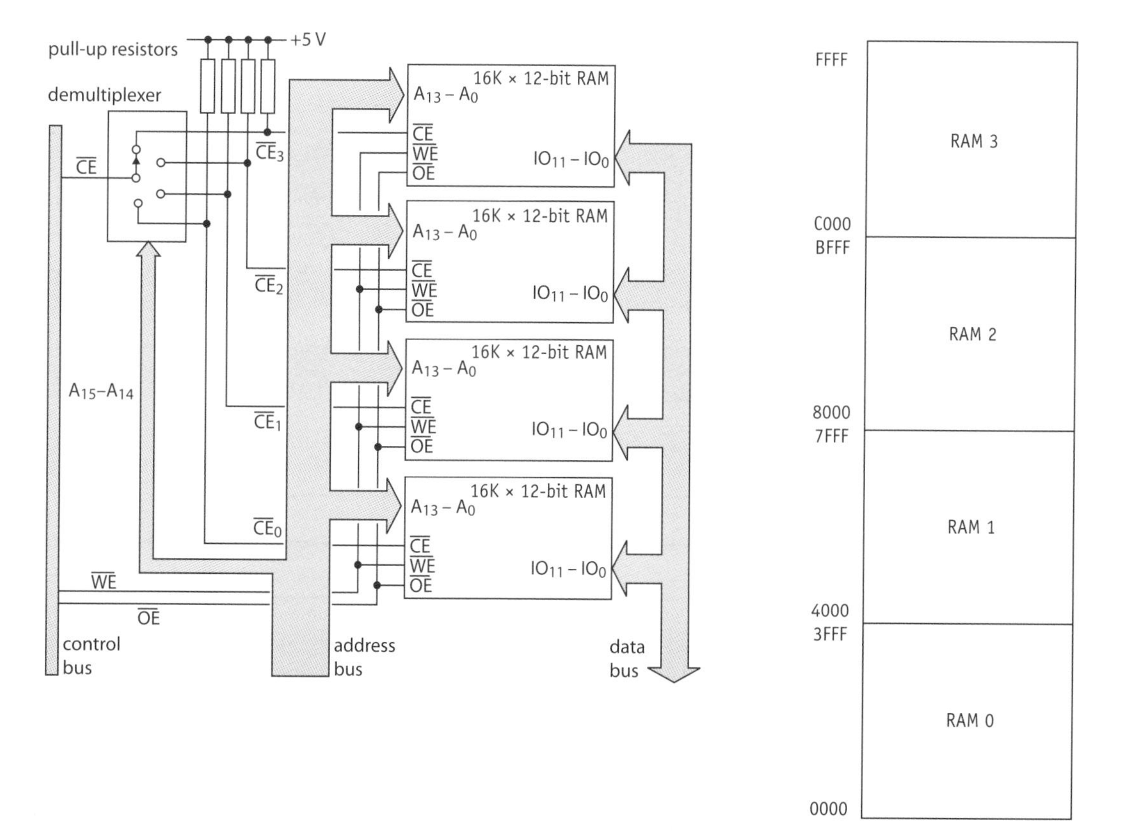

Fig 2.7 Four 16K memories making a single 64K RAM

Fig 2.8 Memory map for the 64K RAM

2.2 Memory cells

There is a bewildering range of different technologies used to make memories, many of them outdated. This section will introduce you to just two. Both are widely used, but neither of them is perfect for every application. The first one is called **static random access memory** (SRAM), and is made from flip-flops and tristates. It has a short access time, and you can write to or read from each location in any order you like, but it forgets everything as soon as its power supply is removed. It also uses a lot of transistors to store each bit, so is expensive. The second type is called **flash memory**, the type used in USB memory sticks and phone memory cards. It only uses a single MOSFET to store each bit, so is relatively cheap. Furthermore, it continues to store information in the absence of a power supply. However, it has a relatively long access time, particularly for write operations, so is no use for systems (like computers) that need fast access to data.

Single cell

It should be fairly obvious from your work at AS that a flip-flop is the ideal component for storing a single bit of information. So the appearance of one at the heart of the memory cell shown in Fig 2.9 should come as no surprise.

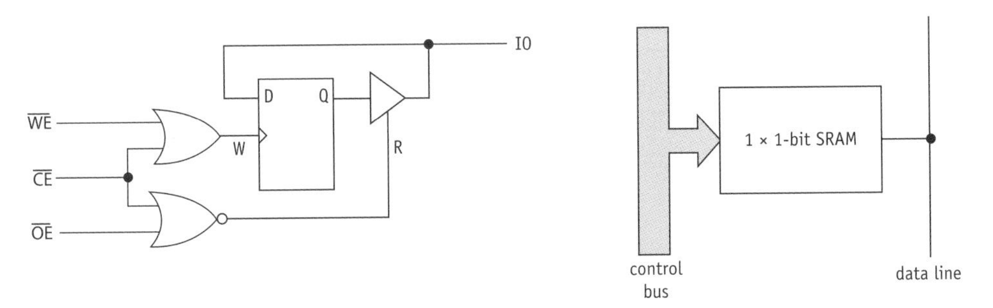

Fig 2.9 A single memory cell

The logic gates and the tristate are there to steer signals to and from the data line IO. Let's start with the read operation. As you can see from the truth table, the R signal which enables the tristate is normally low, disconnecting Q from IO. R only goes high when $\overline{CE}$ and $\overline{OE}$ are low, connecting Q directly to the data line, allowing the bit stored by the flip-flop to control the state of IO.

$\overline{CE}$	$\overline{OE}$	$R = \overline{\overline{CE} + \overline{OE}}$
0	0	1
0	1	0
1	0	0
1	1	0

The write operation is more complicated. Anything placed at the D input is frozen at Q every time a rising edge enters the clock terminal at W. Inspection of the truth table on page 29 shows that this will happen when $\overline{CE}$ is low and when $\overline{WE}$ goes from low to high – just as it does for a memory module.

$\overline{CE}$	$\overline{WE}$	$W = \overline{CE} + \overline{WE}$
0	0	0
0	1	1
1	0	1
1	1	1

More bits

The arrangement of Fig 2.9 is easily extended to handle words rather than just bits. For example, the circuit of Fig 2.10 can read from, or write to, a 4-bit data bus. Each flip-flop stores a different bit of the 4-bit word (or nibble). A nibble placed on the data bus will be latched onto the flip-flops whenever $\overline{CE}$ is low and $\overline{WE}$ rises at the end of a low pulse. It can be subsequently read by holding both $\overline{CE}$ and $\overline{OE}$ low, enabling the four tristates between the flip-flop outputs and the data bus. Note that the tristates are only needed at the output of each flip-flop. The input draws hardly any current, so isn't going to affect the state of the data bus.

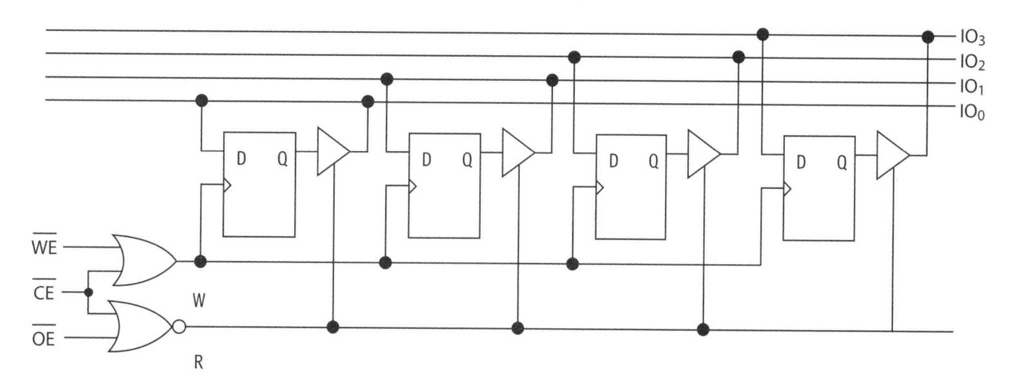

Fig 2.10 A 1×4-bit SRAM

More locations

A useful memory stores more than just one word. Demultiplexers can use the contents of the address bus to route the $\overline{CE}$ signal to just one array of flip-flops, allowing its word to be read from or overwritten, depending on the state of the other two bits of the control bus. Fig 2.11 shows how this can be used to combine four 1×4-bit SRAMs to make a single 4×4-bit SRAM. Of course, the same technique can be used over and over to make smaller memories into larger ones.

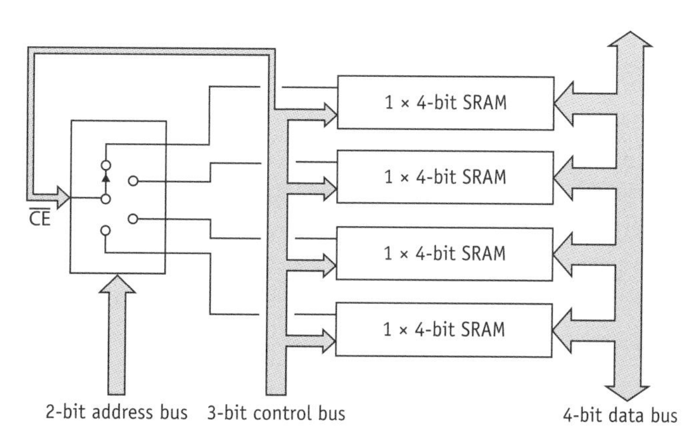

Fig 2.11 A 4×4-bit SRAM

Non-volatile memory

Data stored in a flip-flop is **volatile**. Each time its power supply is turned on, a flip-flop is set or reset at random. Any information that it stores is instantly forgotten the instant that the power supply is turned off. **Non-volatile memory** doesn't have this problem – it only needs a power supply for reading and writing.

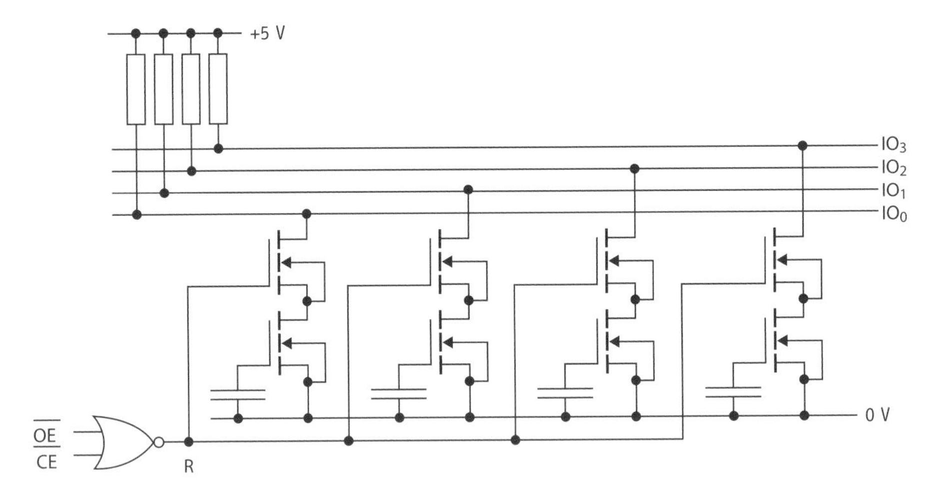

Fig 2.12 A 1×4-bit non-volatile memory

The principle is illustrated in the 1×4-bit ROM (**read-only-memory**) of Fig 2.12. It contains four 1-bit memory cells, each made from a capacitor and a pair of MOSFETs. The top MOSFET is used to **sense** the data stored by the capacitor and the bottom MOSFET. Notice how the top plate of the capacitor is only connected to the gate of the bottom MOSFET. Since there is no current in a MOSFET gate, the capacitor can neither charge nor discharge; its voltage can't change, even when the power supply is disconnected.

Sensing bits

How is the top MOSFET of each pair able to sense the state of the bottom MOSFET? Suppose that the capacitor is charged to +5 V. This forces the bottom MOSFET to its low resistance state. Holding the R line high also puts the top MOSFET in its low resistance state, so it can sink current from one line of the data bus and its pull-up resistor. That current sets up a voltage drop across the pull-up resistor, forcing the line low. So charging up the capacitor allows the cell to store a 0. Conversely, leaving the capacitor uncharged means that the cell stores a 1: the bottom MOSFET will stay in its high resistance state, so there can be no current in the pull-up resistor when R goes high, leaving the data line at +5 V.

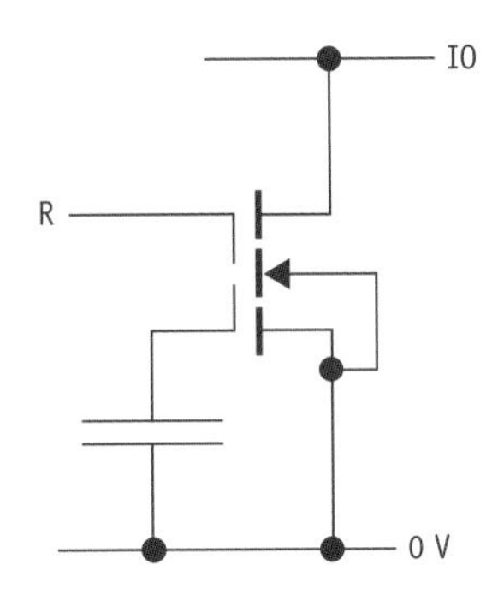

Fig 2.13 One cell of a flash memory

Two gates

Flash memory collapses both MOSFETs into a single one with two gates, as shown in Fig 2.13. Both gates need to be held high to put the MOSFET into its low resistance state, allowing the drain to sink current from the data line IO. So only if the capacitor is charged, when R goes high, will the MOSFET conduct. The layer of insulation between the two gates is very thin. Holding R well above +5 V allows charge to leak through the insulation onto the capacitor, charging it up. Similarly, the capacitor can be discharged by holding R well below 0 V.

2.3 Processing words

My USB memory stick is a good example of a memory. All it can do is store words, only coming to life when it is made part of a larger system. So only when I plug it into my laptop can those millions of bytes be processed by another sub-system called a **central processing unit** (or **CPU**).

CPU structure

Fig 2.14 shows the overall structure of a simple CPU which can combine pairs of bytes in four different ways. At the top is a data bus which carries bytes from the memory (not shown) to one of three **registers**, labelled I, A and B. A register is effectively a 1×8-bit memory with a separate input and output bus (more details overleaf). In the middle are four **processor** modules and a demultiplexer. The demultiplexer directs the **process enable** (or PE) signal to just one of the processors, depending on the byte stored in the **instruction register**. Each processor takes in two bytes and combines them to make just one, according to the table.

I_1I_0	selected processor	processor function
00	A AND B	pairs of bits in bytes A and B pass through an AND gate
01	A EOR B	pairs of bits in bytes A and B pass through an EOR gate
10	A ADD B	byte B is added numerically to byte A
11	A SUB B	byte B is subtracted numerically from byte A

This is how the whole system operates.

- Three separate bytes are loaded, one after the other, from the memory into the I, A and B registers. This places pairs of bytes at the inputs of the processors and sets up the demultiplexer to feed PE to just one of them.
- PE is pulsed high. This activates tristates at the outputs of just one processor, allowing a byte to be fed into register Q.
- While PE is high, a clock pulse into Q latches the byte from the selected processor. That byte is subsequently stored in memory via the data bus.

Notice that the control bus which supplies clock signals for the four registers has been omitted. This keeps the diagram simple and uncluttered.

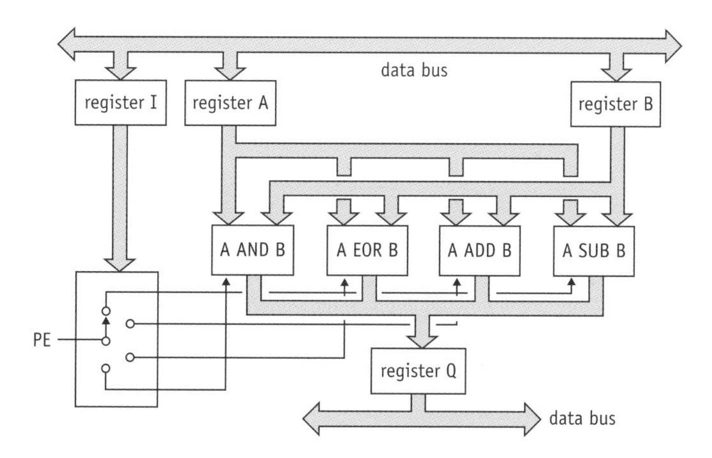

Fig 2.14 A CPU with four different functions

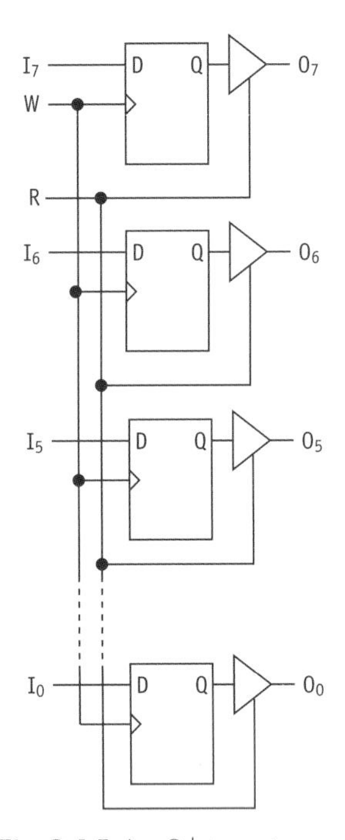

Fig 2.15 An 8-bit register

Registers

You will recall from your work at AS that a register is part of a microcontroller system which can hold bytes ready for processing. It is effectively a 1×8-bit SRAM with separate input and output buses. Fig 2.15 shows how you can make one from D flip-flops and tristates.

You **write** to the register as follows.

- Place the byte at the eight input terminals $I_7 - I_0$.
- Pulse the W line of the control bus high. This latches the byte into the eight flip-flops.

To **read** the byte from the register, just hold the R line high. This opens up the tristates at the output of each flip-flop, allowing the byte through to the data bus.

AND processor

The CPU shown in Fig 2.14 (overleaf) performs four useful functions for a microcontroller (the subject of Chapter 4.) The first of these is the AND function which can be used to selectively reset bits of a word to 0. As you will find out below, this allows you determine the state of just one bit of a word. Fig 2.16 shows you how a bank of eight AND gates and tristates can be used to make the AND processor.

Each gate processes a pair of bits, one from each register. The truth table reminds you what the top gate does when PE is high.

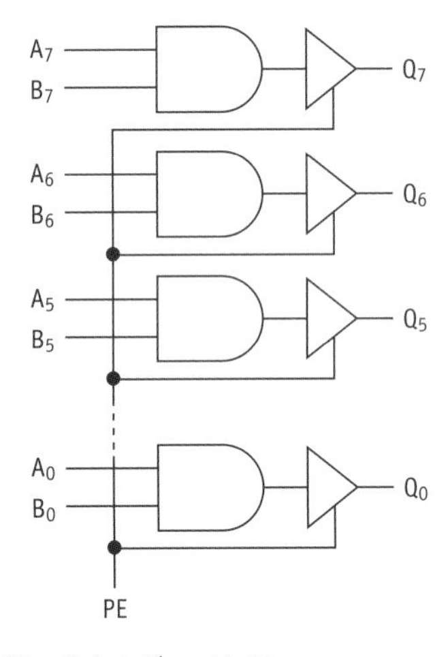

Fig 2.16 The AND processor

B_7	A_7	Q_7
0	0	0
0	1	0
1	0	0
1	1	1

B_7 can be set to 1 or reset to 0. If it is 0, as in the first two rows of the table, then Q_7 will be 0 regardless of the state of A. However, if B_7 is 1 then Q_7 is the same as A_7 – look at the last two rows of the table. So if the B register holds the byte 0010 0001, then the Q register will receive the byte $00A_50\ 000A_0$ when PE is pulsed high. This operation is called **masking**, as it allows you to selectively reset bits of a word to 0. As you will find out in Chapter 4, this is particularly useful to determine the state of a single pin of a microcontroller's input port.

EOR processor

The EOR processor of Fig 2.17 looks very similar to the AND processor on the previous page. It is very useful for selectively changing the state of a single bit of a word.

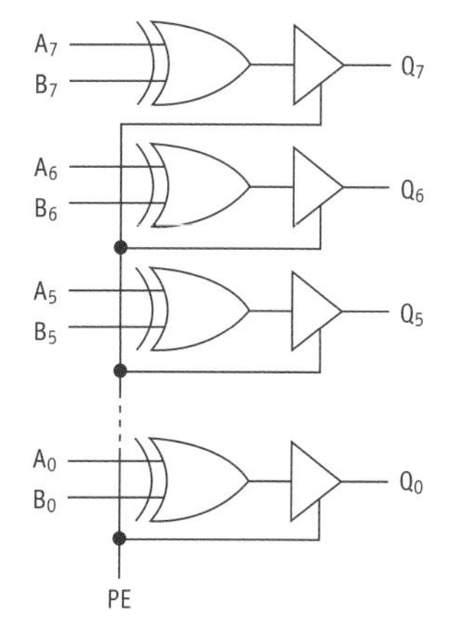

Fig 2.17 The EOR processor

B_7	A_7	Q_7
0	0	0
0	1	1
1	0	1
1	1	0

Suppose that B_7 is reset to 0. Then by inspecting the first two rows of the truth table, you can see that Q_7 is the same as A_7. The bit from register A is not changed as it passes through the processor to register Q. However, if B is set to 1, then Q_7 is the opposite of A_7 – use the last two rows of the truth table to convince yourself. So if the B register holds the byte 0100 1001, then the Q register will receive the byte $A_7\overline{A_6}A_5A_4\overline{A_3}A_2A_1\overline{A_0}$ when PE is pulsed high. This operation is called **toggling**, as it allows you to selectively invert bits of a word. Toggling allows you to change the state of a microcontroller's output port one bit at a time.

Positive numbers

The other two processor modules of Fig 2.14 perform arithmetic operations on the contents of registers A and B. Before you find out how this can be done with logic gates, you need to know how positive and negative numbers can be represented in binary. You may recall from AS how positive numbers can be represented in binary with **binary coded decimal** (or **BCD**). The word DCBA in binary becomes 8D + 4C + 2B + A in decimal. Now consider adding 5 and 1 in BCD, as shown below.

$$\begin{array}{r} 0101 \\ +\ 0001 \\ \hline 0110 \\ \hline 1 \end{array}$$

The first two rows show the numbers to be added, in BCD. The third row shows the result of the addition (the **sum**). Any bits which need to be carried forward are shown in the bottom **carry** row. Starting from the **least significant bit** (or **lsb)** in the rightmost column, 1 + 1 = 2. This is 10 in BCD, so the sum is 0 and 1 is carried forward to the next column. That column now has three bits to be added: 0 + 0 + 1 = 1. The sum is 1 and the carry forward is 0. The final two columns are straightforward: 1 + 0 = 1 and 0 + 0 = 0. The final word in the third row is 0110 = 8×0 + 4×1 + 2×1 +1×0 = 6 – exactly as you would expect.

Negative numbers

Here is the recipe for obtaining the binary representation of a negative number in **two's complement**.

- Write out the BCD representation of the positive number.
- Add a leading zero if the most significant bit is not already zero.
- Invert all the bits.
- Add one.

For example, suppose you needed to represent −5. Write down 0101 to represent +5. Check that the msb is 0. Invert all the bits to obtain 1010. Then add one.

$$\begin{array}{r} 1010 \\ +\ 0001 \\ \hline 1011 \\ \hline \end{array}$$

The result is 1011. You can check that this represents −5 by adding it to +5, as shown below.

$$\begin{array}{r} 1011 \\ +\ 0101 \\ \hline 0000 \\ \hline 111\ \end{array}$$

As you might expect, the answer is zero. The table below shows the two's complement representation of the decimal numbers between +7 and −8. Note that if the msb of the word is a 1 it represents a negative number.

decimal	two's complement
+7	0111
+6	0110
+5	0101
+4	0100
+3	0011
+2	0010
+1	0001
0	0000
−1	1111
−2	1110
−3	1101
−4	1100
−5	1011
−6	1010
−7	1001
−8	1000

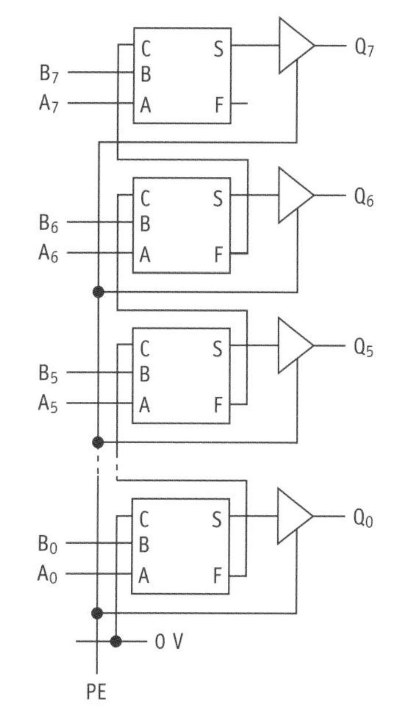

Fig 2.18 Addition processor

Addition processor

Look at the addition processor shown in Fig 2.18. The processing is done by sub-systems called **full adders** with three inputs (A, B and C) and two outputs (S and F). Notice how the C input of each full adder comes from the F output of the full adder below. This is to accommodate the result of $1 + 1 = 0$ carry 1. (The full adder which deals with the two lsb of A and B doesn't have this complication, so its C input is a constant 0.) A full adder obeys this truth table.

C	B	A	S	F
0	0	0	0	0
0	0	1	1	0
0	1	0	1	0
0	1	1	0	1
1	0	0	1	0
1	0	1	0	1
1	1	0	0	1
1	1	1	1	1

You can verify for yourself that the outputs are given by these expressions.

$S = \overline{C}.\overline{B}.A + \overline{C}.B.\overline{A} + C.\overline{B}.\overline{A} + C.B.A$

$F = \overline{C}.B.A + C.\overline{B}.A + C.B.\overline{A} + C.B.A$

With a bit of effort, the expressions for S and F can be recast as follows.

$S = \overline{C}.(\overline{B}.A + B.\overline{A}) + C.\overline{(\overline{B}.A + B.\overline{A})}$

$F = B.A + C.A + C.B$

Fig 2.19 shows how a full adder can be assembled from EOR, AND and OR gates. Note the symmetry of the circuit, with A, B and C being interchangeable.

Fig 2.19 Full adder

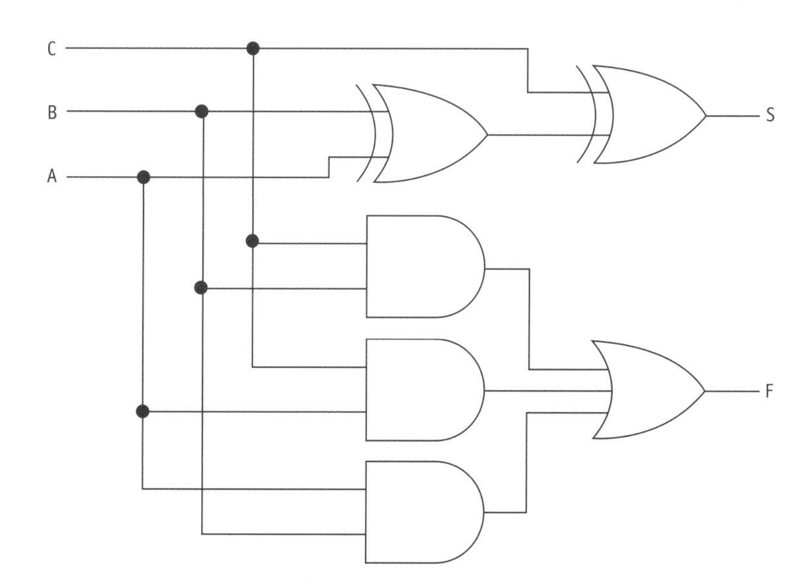

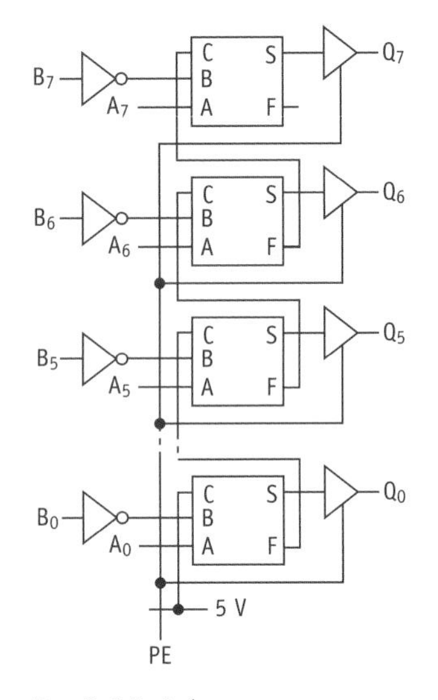

Fig 2.20 Subtraction processor

Subtraction processor

A simple modification turns an addition processor into a subtraction one. This is because the operation A − B is the same as A + (−B). You will recall that in two's complement you multiply a number by −1 by inverting all of the bits then adding one. The inverting is easily done with NOT gates, as shown in Fig 2.20. The addition of 1 is more subtle. Look at the full adder which deals with the two lsbs, A_0 and B_0. Its C input is connected to +5V, effectively adding one all the time.

Single byte processors

Fig 2.21 shows a CPU which performs four useful operations on single bytes stored in register A.

- INC adds one to the number, **incrementing** it.
- DEC subtracts one from the number, **decrementing** it.
- SHL **shifts** each bit one place to the left, effectively doubling the number.
- SHR **shifts** each bit one place to the right, effectively halving the number.

You will meet all of these again in Chapter 4, where you will find out how they can be used in a microcontroller.

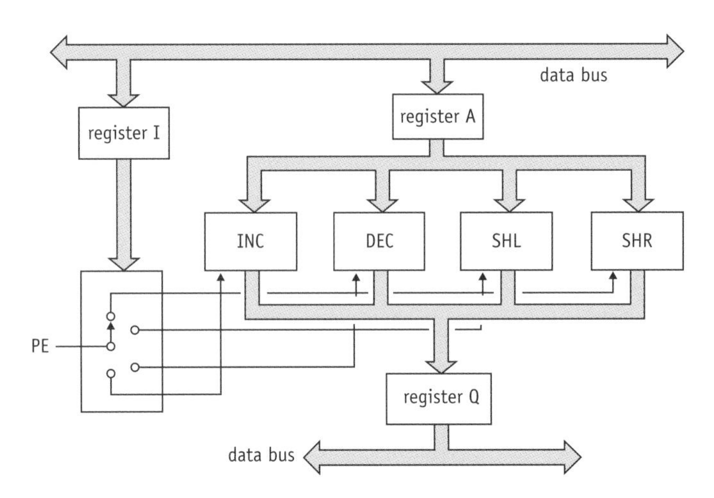

Fig 2.21 Single byte operation CPU

Questions

2.1 Memory modules

1 A memory chip is connected to a data bus, a control bus and an address bus.

(a) What is a bus?

(b) State the function of

(i) the data bus

(ii) the address bus

(iii) the control bus

(c) In what way is the data bus different from the control and address buses?

2 The control bus of a memory chip consists of active-low read, write and enable lines.

(a) Explain the meaning of the term **active-low**.

(b) Describe the sequence of signals required on the control bus to:

(i) read a word from the memory

(ii) write a word to the memory

3 A particular memory chip can hold 2048 different four-bit words.

(a) State the number of lines required for:

(i) the data bus

(ii) the control bus

(iii) the address bus

(b) Draw a diagram to show how four of these chips and a demultiplexer, can be used to make a single 4K×8-bit RAM.

2.2 Memory cells

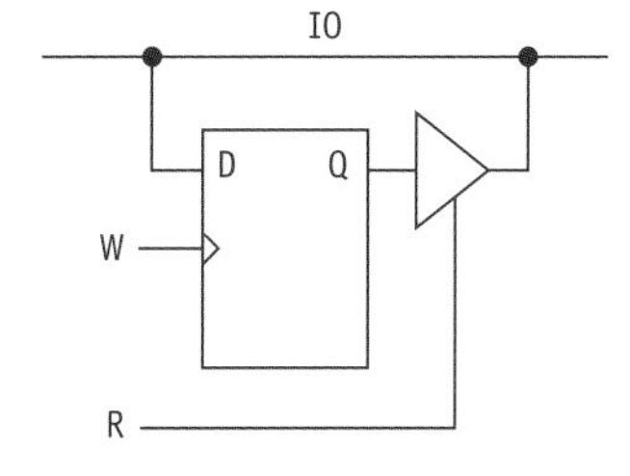

Fig 2.22 Circuit for question 1

1 Fig 2.22 shows a memory cell that can store one bit of information.

(a) Describe the behaviour of the flip-flop.

(b) Describe the behaviour of the tristate.

(c) The two control lines are **active-high**. Explain what this means.

(d) Draw timing diagrams for IO, Q, W and R to show the process of:

(i) reading a bit from the cell

(ii) writing a bit to the cell

(e) The system of Fig 2.22 is an example of volatile memory. Explain what this means.

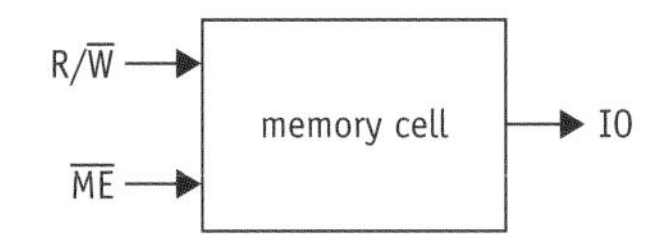

Fig 2.23 Circuit for question 2

2 Fig 2.23 shows a non-volatile memory cell which can store one bit of information.

(a) What is meant by the term **non-volatile**?

(b) The table shows the effect of the control lines on the memory cell.

$\overline{ME}$	R/$\overline{W}$	effect on memory cell
0	0	the bit on IO is stored as $\overline{ME}$ goes high again
0	1	the stored bit is placed on IO
1	0	IO is open-circuit
1	1	IO is open-circuit

Draw timing diagrams for IO, $\overline{ME}$ and R/$\overline{W}$ to show how to:

(i) read a bit from the cell

(ii) write a bit to the cell

(c) Draw a diagram to show how the cells could be combined to make a 1 × 3-bit memory with a two-bit control bus.

(d) Draw a diagram to show how two 1 × 3-bit memories could be combined to make a single 2 × 3-bit memory with a two-bit control bus.

(e) How many of the cells of Fig 2.23 would be required to make a 64K × 8-bit memory module?

3 This question is about the non-volatile memory cell of Fig 2.24.

(a) State and explain the state of O when R is low.

(b) State and explain the state of O when R is high and:

(i) G is high

(ii) G is low

(c) Explain why the circuit of Fig 2.24 is a **read-only non-volatile memory.**

(d) Show how the addition of a tristate and logic gates can make the memory cell of Fig 2.24 into a memory cell with a two-bit control bus which obeys this table. Justify your circuit.

$\overline{CE}$	$\overline{OE}$	effect on memory cell
0	0	stored bit placed on output line
0	1	output line high
1	0	output line open-circuit
1	1	output line open-circuit

(e) Explain why the cell you have drawn for (d) must have a tristate output.

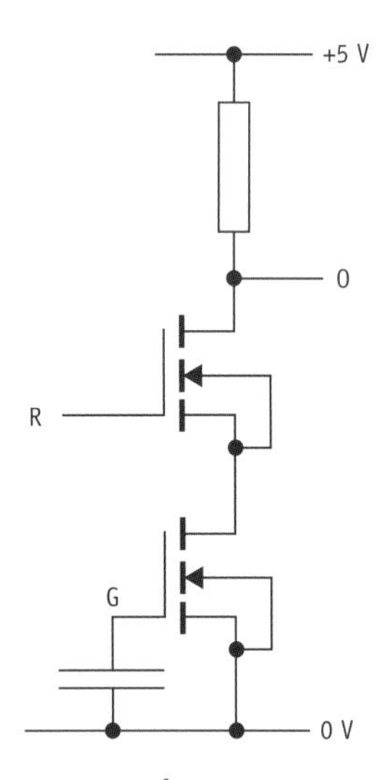

Fig 2.24 Circuit for question 3

2.3 Processing words

These questions are about the processing system of Fig 2.25. Each register holds a 4-bit word.

1 Register A can latch a 4-bit word from the data bus and pass it directly to the processor.

(a) Show how register A can be assembled from flip-flops. Label the input, output and control lines. Explain the operation of the system.

(b) The processor is required to combine the four bits of A and B when PE goes high, as shown in the truth table.

A_n	B_n	C_n
0	0	0
0	1	0
1	0	1
1	1	0

Show how the processor can be assembled from logic gates and tristates.

(c) Why does register C need tristate outputs?

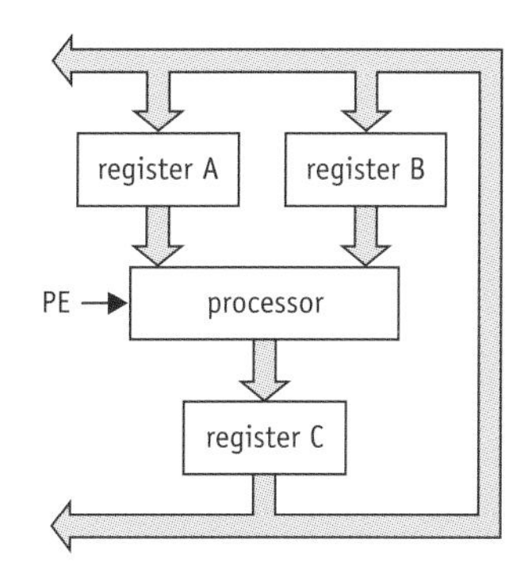

Fig 2.25 Circuit for questions 1, 2 and 3

2 For this question, the processor of Fig 2.25 adds the numbers in A and B and places the answer in C.

(a) If A contains +4 and B contains −5, write down the contents of the registers in binary code.

(b) Write down the contents of C in binary code when PE is pulsed high.

(c) Show how the processor can be made from full adders and tristates.

(d) Write out a truth table for the full adder. Show how it can be assembled from NAND gates. Use Boolean algebra to justify your circuit.

3 For this question, registers A and B contain 0101 and 1100 respectively. Explain the word passed to register C when the processor is made from

(a) AND gates

(b) EOR gates

(c) OR gates

Learning summary

By the end of this chapter you should be able to:

- draw a block diagram for a memory module, with its buses
- explain how to write and read words to and from a memory
- show how to make a volatile memory cell from a flip-flop and a tristate
- show how to make a non-volatile memory cell from a capacitor and a MOSFET
- combine small memories to make larger ones
- use two's complement coding to represent numbers
- understand the use of logic gates to combine words stored in registers

CHAPTER 3 Servo control

3.1 On-off control

Don't try this on your own – you could hurt yourself: Stand on one leg for a few seconds. You shouldn't find it difficult to stay upright. Now close your eyes, still only standing on one leg. If you are young, you might be able to stay upright for a minute or two. I fall over after a few seconds. This is because closing your eyes removes the main source of feedback for your body, so that it can no longer accurately assess your state of uprightness. This chapter is going to show you how electronic control systems (called **servo loops**) use feedback to maintain control of physical quantities such as temperature, position, speed, light level and voltage.

Temperature control

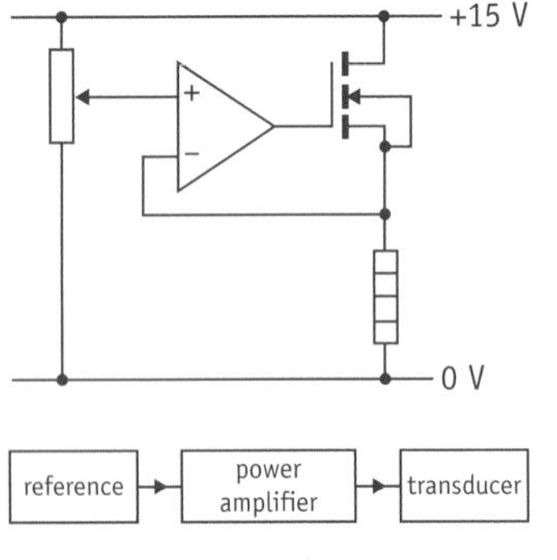

Fig 3.1 An open-loop temperature control system

Suppose you needed to keep an enclosure at the same temperature all of the time. It might be an incubator for eggs, or a cot in a premature baby unit. The simplest approach is to use a system like the one shown in Fig 3.1.

Consider what happens when the system is switched on. Energy pours out of the heater, raising the temperature of the enclosure. The temperature doesn't go up for ever because the rate of energy loss from the enclosure increases with its temperature. So the temperature eventually levels off when energy is being lost as fast as it is being gained. The power of the heater is determined by the setting of the potentiometer, so the position of the wiper should effectively set the final equilibrium temperature of the enclosure. However, suppose that the temperature outside the enclosure falls. This increases the rate of energy loss, lowering the equilibrium temperature. Without a feedback loop, the system would be unaware of this change and therefore unable to compensate for it.

Open-loop

The **control system** of Fig 3.1 has three distinct subsystems.

- The potentiometer is the **reference**. It provides a signal which ultimately determines the value of the physical quantity being controlled.
- The op-amp and MOSFET are arranged as a voltage follower, making a **power amplifier** which passes on the voltage from the wiper but allows a much larger current.
- The heater is a **transducer**. It uses electrical energy to change some physical quantity – temperature in this case.

Closed-loop

The circuit of Fig 3.2 does a better job of controlling temperature. The thermistor is inside the enclosure, so that the **sensor** signal S which enters the non-inverting input of the op-amp contains information about the temperature of the enclosure (increasing temperature results in increasing voltage at S). The op-amp then switches the heater on or off, depending on how S compares with the reference signal R. The presence of this **feedback loop** means that this is a **closed-loop system**, allowing it to automatically and continuously adjust itself to achieve the desired outcome set by the reference signal. Notice that the transducer is controlled by a **switch** instead of a power amplifier. This often makes more efficient use of electrical energy, resulting in less heating of the MOSFET.

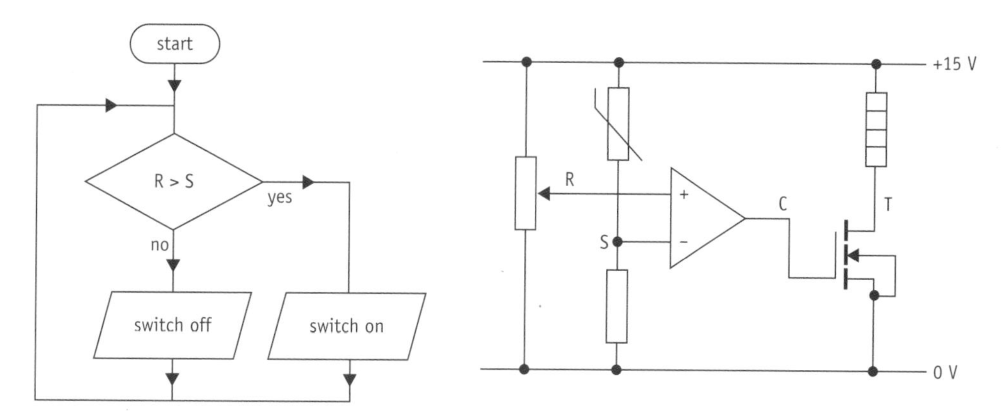

Fig 3.2 Closed-loop temperature control circuit with its system flowchart

Study the systems flowchart of Fig 3.2. It shows that the op-amp **comparator** is continuously assessing which of two signals (R or S) is the greater and then switching the heater on or off accordingly.

Negative feedback

The block diagram of Fig 3.3 shows the feedback loop as a dotted line. The dots show that the information is not transferred from the transducer to the sensor in electrical form. Notice that the dotted feedback loop allows the system to reduce the difference between S and R – hence the name **negative feedback**.

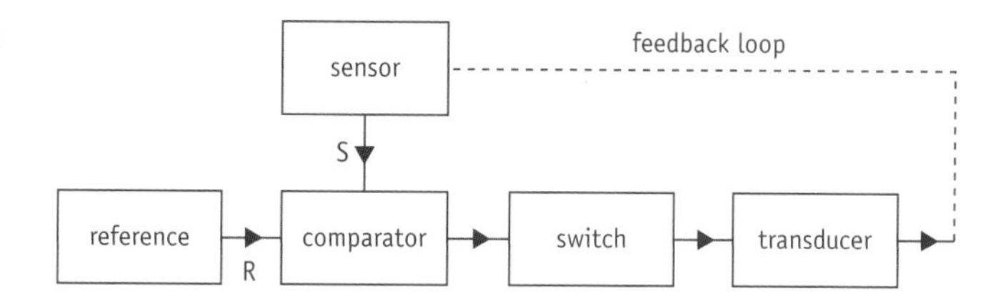

Fig 3.3 Block diagram for an on-off servo control system

On-off response

The graphs of Fig 3.4 show how the signals in the system change with time.

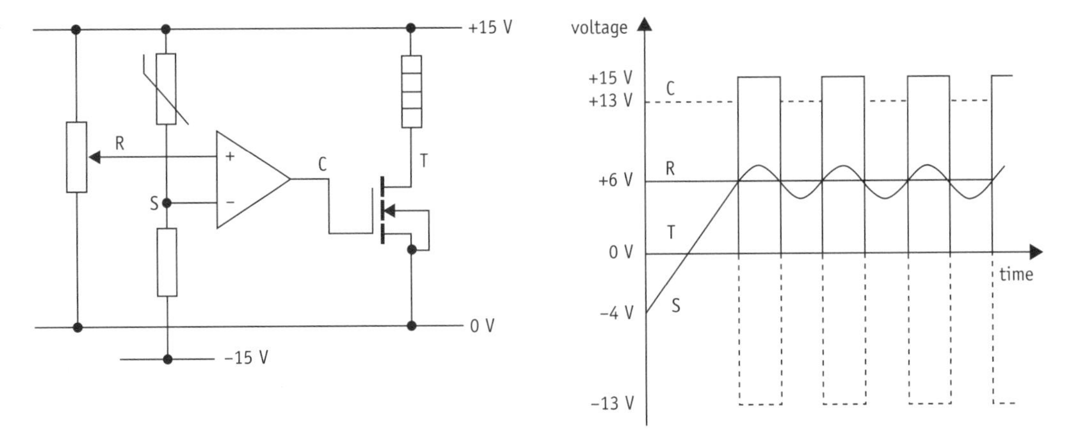

Fig 3.4 The temperature never settles to a constant value

Four graphs are shown on a single set of axes. The reference signal R is set to an arbitrary but constant value of +6 V. The sensor signal starts off at −4 V, showing that the temperature of the enclosure is too low. The output of the comparator C duly saturates at +13 V, switching on the MOSFET, allowing a current in the heater which forces the signal at T to drop to almost 0 V. So what happens next?

- The enclosure heats up, increasing the value of S.
- When S goes above R, the comparator signal C saturates instantly at −13 V.
- The MOSFET switches off, and T rises to +15 V.

Notice that even when the heater is turned off, the temperature of the enclosure – as measured by the signal S – continues to rise for a while. This is because energy still pours out of the heater when it is switched off and it takes time for that energy to reach the thermistor. We say that the system has **inertia**: it cannot respond instantly to changes. However, energy is still leaking out of the enclosure, so it is only a matter of time before the temperature starts to drop.

- The enclosure cools down, decreasing the value of S.
- When S goes below R the comparator signal C saturates instantly at +13 V.
- The MOSFET switches on, sourcing current, and T falls to almost 0 V.

Once again, the system's inertia means that there is a time delay between the heater going on and the value of S rising back up to the value of R.

Overshoot

It should be obvious by now that using on-off feedback can only control the average temperature of the enclosure. At any instant, the actual temperature, as measured by the signal S, is likely to be above or below the desired temperature as set by the reference signal R. The temperature never settles down; it **hunts**, oscillating above and below its average value. This is often perfectly acceptable for heating houses, usually because the considerable inertia of the system results in a long period for the hunting oscillations.

Light control

Although hunting may often be acceptable for temperature control, it is rarely so for light control. Consider the circuit shown in Fig 3.5.

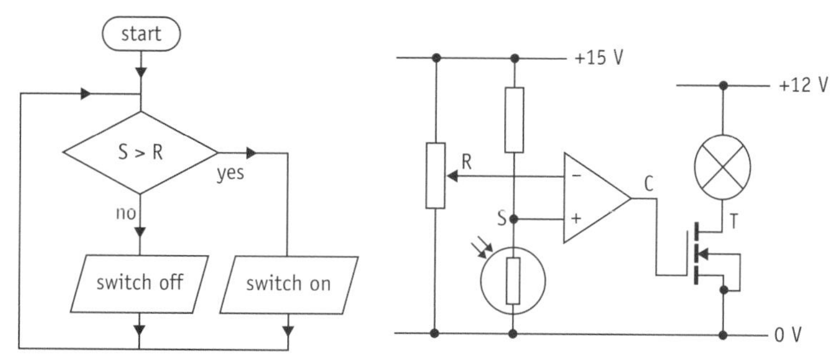

Fig 3.5 On-off light level control

The system has the usual five blocks of an on-off feedback system:

- A potentiometer to set up the reference signal R.
- An LDR and pull-up resistor to form the sensor; the voltage at S increases as it gets darker.
- An op-amp to act as a comparator; C saturates at +13V or −13 V, depending on the voltages at S and R.
- A MOSFET as the switch.
- A lamp as the transducer which produces light.

As you can see from the systems flowchart, when there is enough light around, the lamp stays switched off. But what happens when it gets dark? S rises above R, C saturates at +13 V and the lamp switches on. Almost immediately, this floods the LDR with light, forcing S well below R again. So C saturates at −13 V and the lamp switches off again, leaving the LDR in the dark once more, etc, etc. . .

Overall response

It should be obvious that when it gets dark, the lamp will flicker as it switches on and off rapidly. As long as light from the lamp can reach the LDR, there is no way that the lamp can stay on continuously. However, the response time for a lamp can be quite short, so you may not notice the flickering. Even so, since the lamp will spend some of the time on and some of the time off, it will only be at part power. The overall response is summarised in the table, where the setting of R defines the difference between light and dark.

external conditions	system response
light	lamp stays off continually
dark	lamp flickers on and off rapidly

Evidently, on-off control is not a good idea when the system has little inertia, and the short response time is going to result in large amplitude oscillation. You will find out later on how 'proportional feedback' gets around these difficulties.

Speed control

For many applications, on-off control is satisfactory for maintaining a constant speed of rotation for an electric motor. This is because the motor and what it is running usually have a lot of inertia, so each burst of power as the motor switches on doesn't result in a dramatic change of speed. A typical circuit for on-off speed control is shown in Fig 3.6.

Fig 3.6 Speed control system

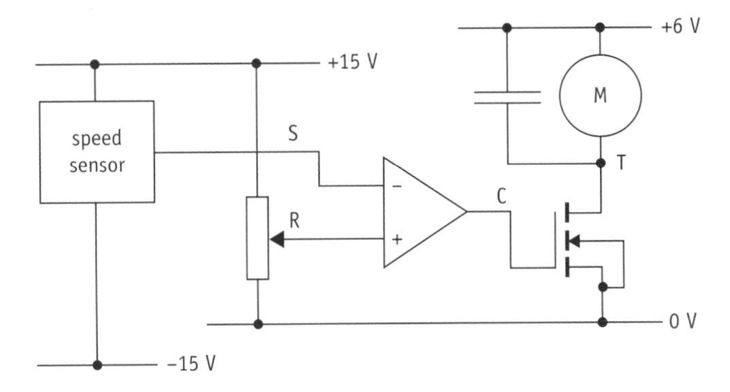

The circuit contains a couple of new elements. Firstly, there is a capacitor in parallel with the motor. This absorbs voltage spikes on the +6 V supply rail caused by the rapid disconnection of the spinning coils inside the motor. Secondly, the speed sensor is shown just as a three-terminal device when it is actually a separate complex sub-system. Otherwise, the system is the same as any other on-off control system. Every time S goes above R, C saturates at −13 V and the switch is turned off. Without any current in it, the motor slows down until S is below R once more, C shoots up to +13 V and the switch is turned on again. The system hunts but, as with the temperature control system, the large inertia of the system makes this acceptable. Short bursts of current in the motor can keep it going at a tolerably steady speed.

+15 V
infrared
LED
22 kΩ
P
c
1 kΩ
e
0 V
phototransistor

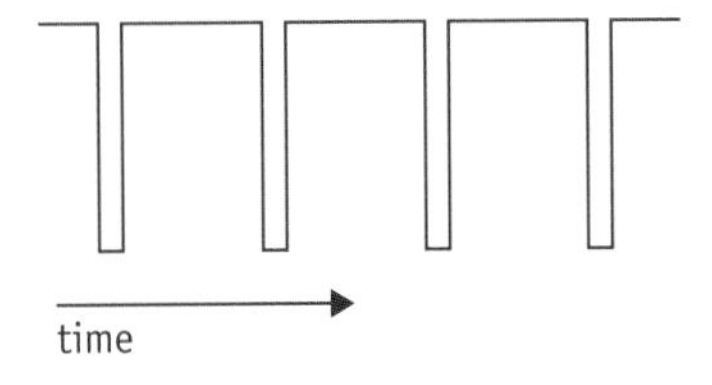

Fig 3.7 Infrared speed detection system

Speed sensor

There are many ways of sensing the speed of a motor shaft. They usually require the production of a fixed number of pulses for each revolution of the shaft. For example, consider the slotted wheel shown in Fig 3.7. The slots pass between the **infrared LED** and **phototransistor** shown in the circuit, allowing through eight short bursts of infrared for each turn of the shaft. Each burst of infrared allows current in the phototransistor from **collector** (c) to **emitter** (e), making the voltage at P drop below +15 V. The resistor between the collector and the top supply rail sets the current required to make P saturate at 0 V.

$I = ?$

$V = 15 - 0 = 15$ V

$R = 22$ kΩ

$$I = \frac{V}{R} = \frac{15}{22 \times 10^3} = 6.8\times10^{-4} \text{ A or } 0.7 \text{ mA}$$

Compare this with the steady 14 mA in the infrared LED (whose forward bias voltage is typically 0.8 V). The **current transfer ratio** therefore needs to be at least 0.7 / 14 = 0.05 (or 5%) to guarantee 15 V pulses at P when the shaft is turned. In practice, transfer ratios of 20 % are easy to obtain by placing the components close to each other.

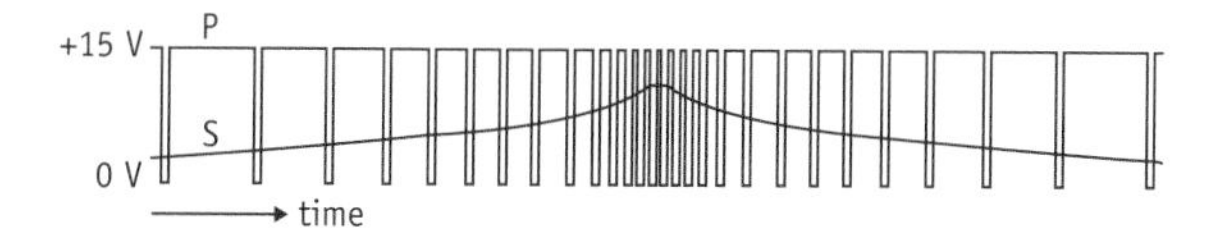

Fig 3.8 As the pulses get closer together, the voltage at S has to rise

Frequency to voltage

Fig 3.8 shows the signal at P as the shaft speeds up and slows down. Each time a slot passes between the LED and the phototransistor, P goes low. The line labelled S is the required service output S of Fig 3.9. A system whose output voltage rises as the frequency of input pulses rises is known as a **frequency-to-voltage converter**. Its block diagram is shown in Fig 3.9.

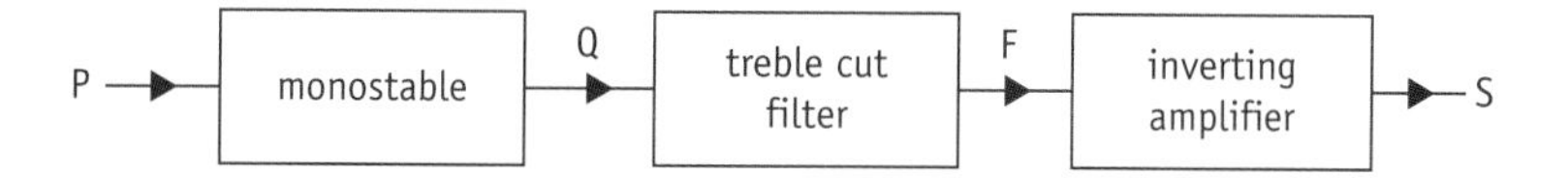

Fig 3.9 Block diagram for a frequency-to-voltage converter

Here is how the system operates.

- The monostable produces a fixed length pulse at Q for each falling edge at P. A pulse length of 1 ms would allow the shaft to rotate at up to $1/8\times10^{-3} = 125$ times a second before the pulses overlap.
- The treble cut filter removes the high frequency components of the signal at Q, leaving the low frequency components that make up S. A break frequency of 10 Hz will give the system a response time of about $1/10 = 0.1$ s, letting through only 1% of the unwanted 1 kHz signals from the 1 ms pulses.
- The inverting amplifier multiplies the signal at F by -1. This is necessary because the filter inverts the signal, and we want S to be positive.

The circuit diagram is shown in Fig 3.10. The NAND gates of the monostable are run off supply rails at +15 V and 0 V to avoid the need for a separate 5 V supply, so the pull-down resistor needs to be at least 30 kΩ instead of the usual 10 kΩ.

$T = 1$ ms

$R = 30$ kΩ

$C = ?$

$$T = 0.7RC$$

$$C = \frac{T}{0.7R} = \frac{1\times10^{-3}}{0.7\times30\times10^{3}} = 4.7\times10^{-8}\text{ F or 47 nF}$$

The break frequency needs to be about 10 Hz.

$f_0 = ?$

$R = 160$ kΩ

$C = 100$ nF

$$f_0 = \frac{1}{2\pi\times160\times10^{3}\times100\times10^{-9}} = 10\text{ Hz}$$

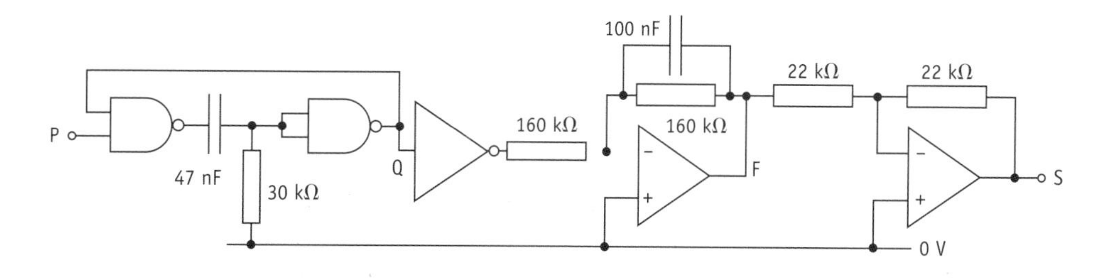

Fig 3.10 Circuit diagram for the frequency-to-voltage converter

Position control

The last on-off control system you need to be aware of is a bit different (and doesn't work very well). The circuit is shown in Fig 3.11.
As with all on-off feedback systems, there are four distinct blocks.

- A **master potentiometer** sets the reference signal R.
- A **slave potentiometer** generates the sensor signal S. The shaft of this potentiometer is connected via gears to some object whose position needs to be controlled. This might be the pen of an X-Y plotter, the rudder of an aeroplane or the cutting tool of a lathe.
- An op-amp compares S and R, so that C saturates at +13 V or −13 V as required.
- A power amplifier instead of a switch controls the current in the motor. This is because the motor shaft must be able to rotate in either direction. So if it turns clockwise when T is positive, it turns anticlockwise when T is negative.

As the motor shaft spins, it changes the position of the object. In turn, this alters the setting of the slave potentiometer. This physical connection between motor and potentiometer is shown as a dotted line in Fig 3.11.

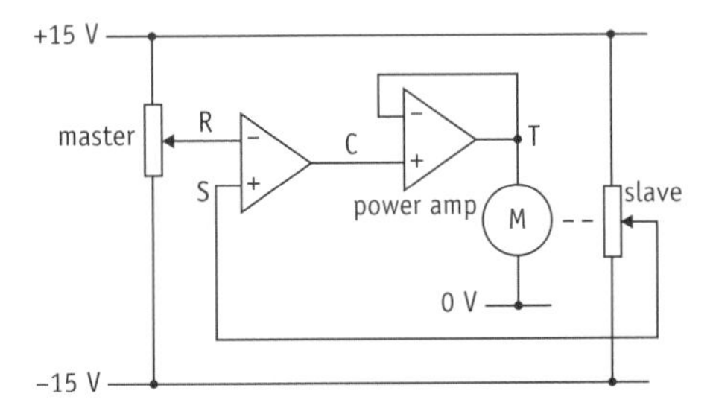

Fig 3.11 Position control system using on-off feedback

Response curve

The graphs of Fig 3.12 illustrate what happens when the setting of the master potentiometer is changed suddenly.

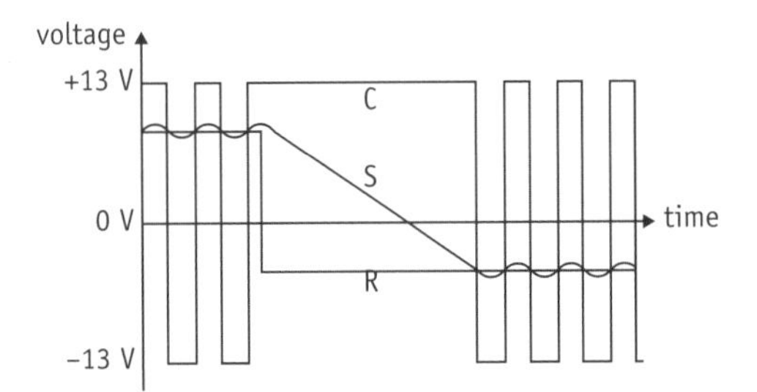

Fig 3.12 Voltage–time graphs for the circuit of Fig 3.11

When the reference signal R suddenly drops from a steady positive value to a steady negative value, the comparator signal C saturates at +13 V. The motor shaft spins clockwise, moving the wiper of the slave potentiometer so that the sensor signal S drops steadily. Once S reaches the new value of R, C saturates at −13V. The inertia of the motor and what it is moving means that it will carry on moving clockwise as it slows down, so S overshoots R. Eventually the shaft starts turning anticlockwise, making S rise once more – until it goes above R and C saturates at +13 V once more. Clearly, the system can never settle down to a steady position as the motor shaft will always be driven clockwise or anticlockwise by the power amplifier. The next section of this chapter will show you a far better way of achieving precise control of position – proportional feedback.

3.2 Proportional control

The secret of precise control of any physical quantity is to make the system respond more slowly as it approaches the desired state. This allows the signal driving the transducer T to change more slowly as S gets closer to R – until T is rock steady when S is the same as R.

Precise position

Fig 3.13 shows the circuit diagram of a position control system which uses proportional feedback.

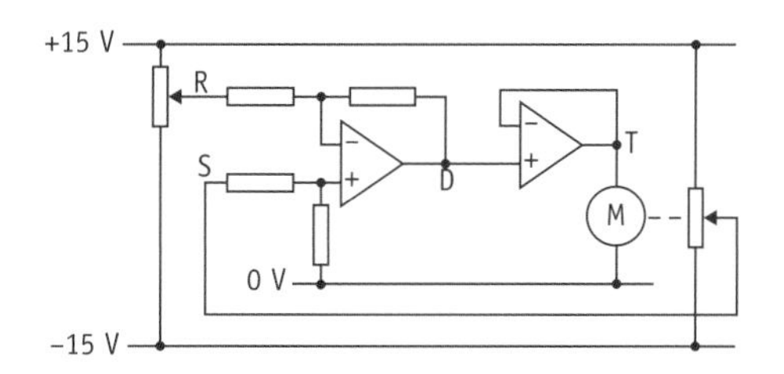

Fig 3.13 Proportional control of position

If you compare it with the on-off circuit of Fig 3.11, you will notice only one difference. The comparator has been replaced with a **difference amplifier**. You will recall that if all four resistors are identical, the output $D = S - R$ provided that D is not saturated. The voltage–time graphs of Fig 3.14 show what effect this one change has on the overall response of the system.

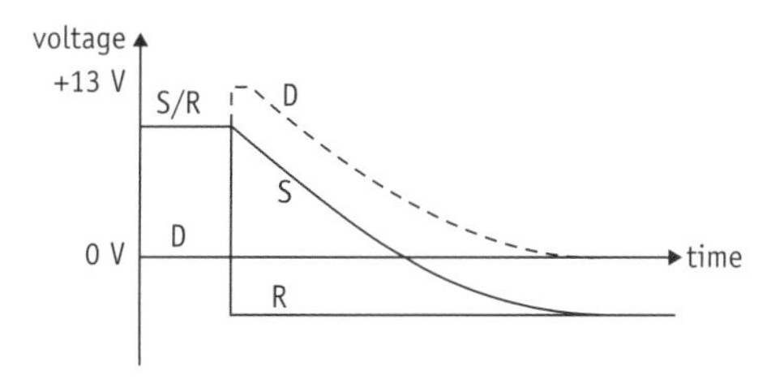

Fig 3.14 Response curves for the circuit of Fig 3.13

Suppose that R is some positive value and that S has the same value, so that D is 0 V. The power amplifier passes this on to T unchanged, so the motor doesn't spin at all. It just stays where it is. Now let the reference signal R suddenly dive to a new steady negative voltage, as shown in Fig 3.14. This is what happens next.

- S is now way above R, so D immediately shoots up to a high positive value. In this case, R has dropped by more than 13 V, so D saturates at +13 V.
- T follows D, putting a large positive voltage across the motor, setting its shaft spinning and moving the wiper of the slave potentiometer.
- The sensor signal S starts to drop towards the new value of R. When $S - R <$ 13 V, D no longer saturates and the voltage across the motor starts to drop.
- A lower voltage at T means that the shaft spins more slowly. So S changes less rapidly than before. Nevertheless, D continues to drop, so the motor shaft moves more and more slowly, until $S - R = 0$ V.

Mission accomplished. The wiper is at its new required position and the shaft has stopped moving completely – until the next change of reference signal R.

Drawbacks

Proportional position control certainly avoids the problem of hunting, but it does have two drawbacks.

- The response of the system to changes of the reference signal R is much slower than it is for an on-off control system. In theory, because the shaft position changes ever more slowly as it approaches its final setting, a proportional control system takes forever to get S the same as R.
- The power of the motor gets smaller and smaller as the shaft approaches its final setting. Eventually, in a real system, friction will stop the motor from moving the shaft, so it will settle to a position a little bit short of where S is the same as R.

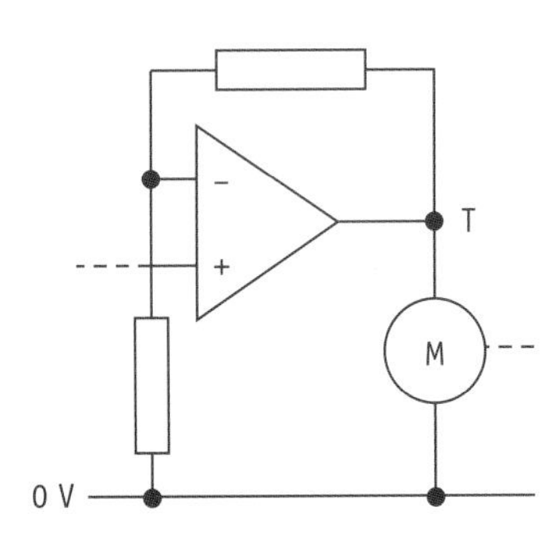

Fig 3.15 Extra gain improves the response time and precision

Both of these drawbacks can be reduced by increasing the gain of the power amplifier, as shown in Fig 3.15. (The extra pair of resistors turn the voltage follower into a non-inverting amplifier.) However, you need to be careful about how much gain you inject into the system – if there is too much, the system will start to behave like an on-off one and hunt instead of settling to a steady final value!

Ramp generator

The secret of every proportional control system is the presence of a sub-system which obeys this table.

input signal	output response
positive	decreases
zero	no change
negative	increases

In the case of position control (Fig 3.15), this sub-system is the motor itself.

- A positive voltage across the motor makes the shaft spin clockwise, moving the wiper of the slave potentiometer downwards.
- No voltage across the motor leaves the wiper where it is.
- A negative voltage across the motor moves the wiper upwards.

For all other proportional control systems, a **ramp generator** (Fig 3.16) needs to be included.

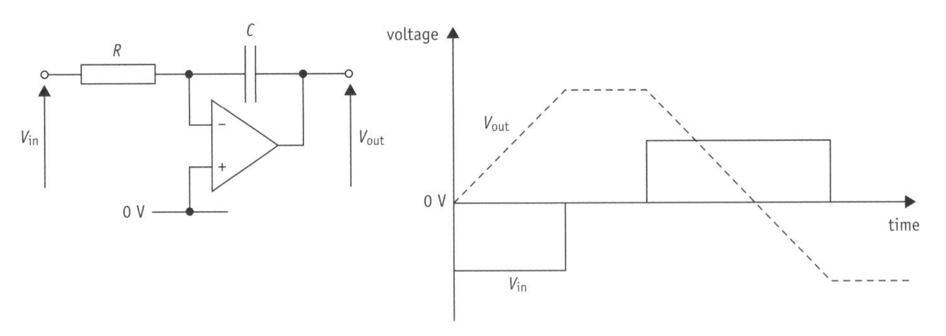

Fig 3.16 Ramp generator

The graph of Fig 3.16 shows how the voltage at the output depends on the voltage at the input.

- The output **ramps up** when the input is negative.
- The output doesn't change when the input is zero.
- The output ramps down when the input is positive.

Of course, this only applmies when the output isn't saturated at +13 V or −13V.

Charging and discharging

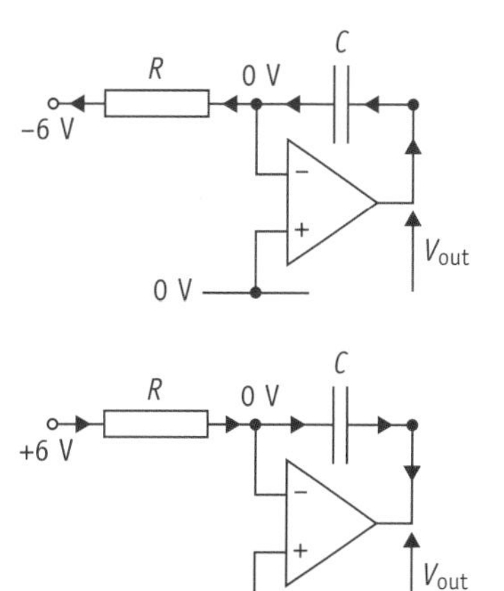

Fig 3.17 The output changes whenever the capacitor charges or discharges

So why does a ramp generator behave like this? Suppose that the voltage at the input is −6 V, as shown at the top of Fig 3.17. Negative feedback requires that the inverting input sits at the same voltage as the non-inverting input, 0 V. So charge flows through the resistor from right to left. That charge comes off the left-hand plate of the capacitor (there is no current at the inputs of an op-amp), so charge must also flow onto the right-hand plate from the op-amp output. As the capacitor charges up, so the voltage across it increases, increasing the value of V_{out} steadily with time. Of course, the output can't go up forever – it will eventually saturate at +13 V.

On the bottom of Fig 3.17, you can see what happens when the input is positive. The resistor sources charge onto the left-hand plate of the capacitor, so the voltage of its right-hand plate has to drop steadily with time as charge flows off it into the output of the op-amp – until it saturates.

Ramp generator calculations

The change of output voltage ΔV_{out} in the time interval Δt can be calculated with this formula.

$$\Delta V_{out} = -\frac{V_{in}\Delta t}{RC}$$

Consider the ramp generator of Fig 3.18. If the output is initially 0 V, how long will it take for the output to saturate when the input is lowered to −5 V?

$\Delta t = ?$

$\Delta V_{out} = 13\text{ V} - 0\text{ V} = +13\text{ V}$

$V_{in} = -5\text{ V}$

$R = 27\text{ k}\Omega$

$C = 470\text{ nF}$

$$\Delta V_{out} = -\frac{V_{in}\Delta t}{RC} \qquad \Delta t = -\frac{\Delta V_{out}RC}{V_{in}}$$

$$\Delta t = -\frac{13 \times 27 \times 10^{3} \times 470 \times 10^{-9}}{-5} =$$

$$\Delta t = 3.3\times10^{-2}\text{ s or 33 ms}$$

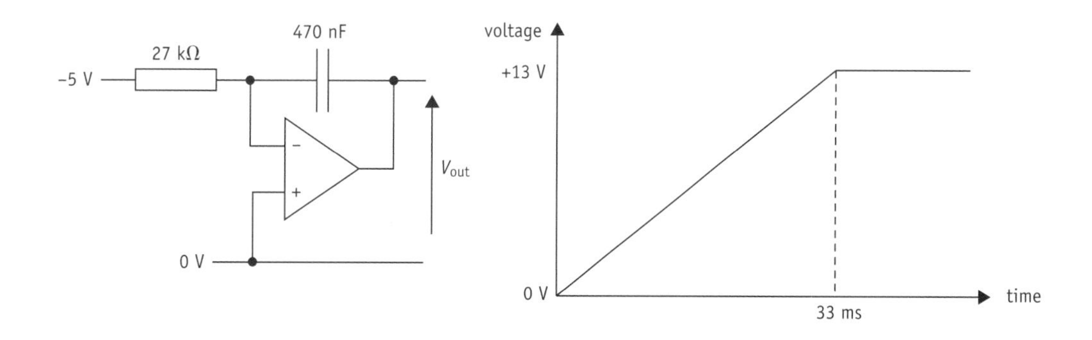

Fig 3.18 A negative input voltage makes the output ramp up

Block diagram

The block diagram of Fig 3.19 applies to many different proportional control systems – including temperature, light level and motor speed.

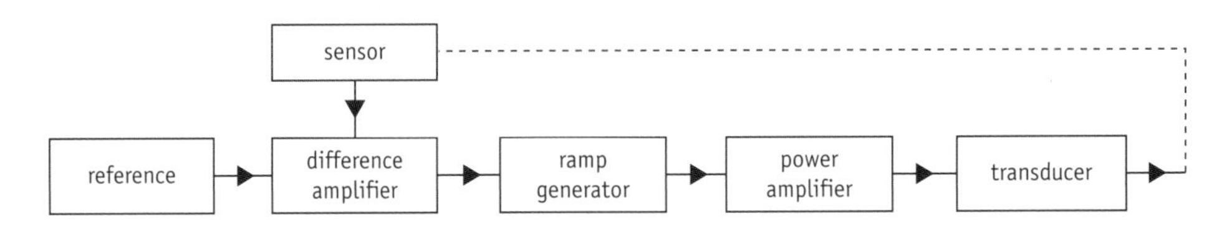

Fig 3.19 Block diagram for a proportional control system

Notice where the ramp generator has been placed in the loop, between the difference and power amplifiers. Any non-zero difference signal will make the input of the power amplifier ramp up or down, changing the voltage across the transducer. When the difference signal reaches zero, the power amplifier input is left at whatever is required to make the transducer do what it has to do.

Temperature control

Fig 3.20 shows a circuit diagram for temperature control.

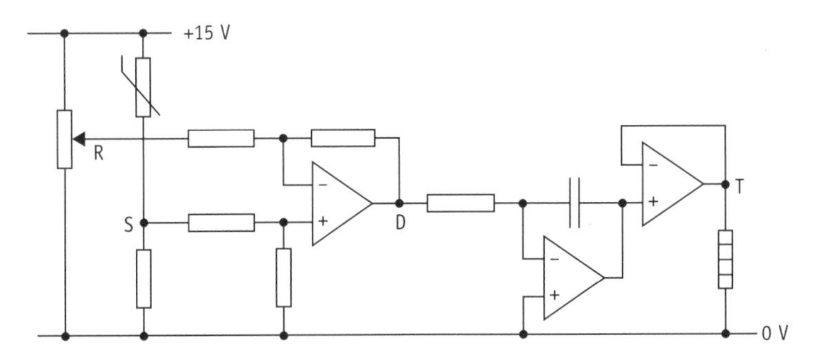

Fig 3.20 Proportional control of temperature

When designing a circuit of this type, you need to pay attention to two things.

1. Is the response time appropriate? If it is too short, the system will hunt. The trick is to make the response time much longer than the period of oscillation of the equivalent on-off control system. The response time is roughly equal to the time constant *RC* of the ramp generator.
2. Is the feedback negative? If it isn't, then the system will end up at one or either end of its range.

You check the type of feedback by showing that the difference signal gets smaller as time goes on. Start off with the system far from its final state, as follows:

- Suppose that R is above S (the system is too cold).
- Then D = S − R will be negative.
- The voltage T across the heater will ramp up, increasing its power.
- The system will heat up, raising the voltage at S.
- The difference signal D = S − R will be smaller than it was.

If this analysis leads to an increase in the difference signal, the cure is easy. Just swap the two signals at the difference amplifier inputs.

3.3 Power supplies

For the purposes of this chapter, a power supply is a control system which maintains a steady low d.c. voltage at its output, using the high voltage a.c. mains supply as its only source of energy. You probably use one regularly to charge up your mobile phone. As well as a servo control system to determine the voltage at the output, a power supply needs to convert a.c. into d.c.

Voltage regulator

Fig 3.21 shows a circuit for a **voltage regulator**, the servo control sub-system in a power supply.

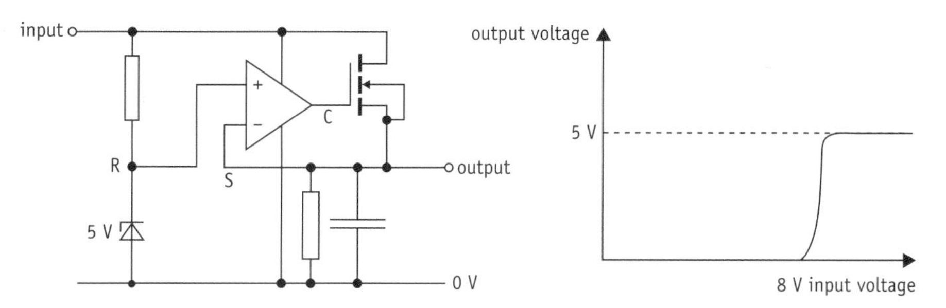

Fig 3.21 Voltage regulator

Notice that the entire circuit derives its power from the signal at the input, with the lower supply rail of the op-amp at 0V. Although this means that the op-amp output can't swing to negative values, this doesn't matter. This is how the system operates.

- The Zener diode and pull-up resistor generate a steady 5 V reference signal R.
- The op-amp compares R with S, the voltage at the output terminal, making C go high or low accordingly.
- The signal at C switches the MOSFET on or off, controlling the current at the output terminal.

Normally, proportional control would be the control system of choice, but a voltage regulator uses on-off control. You can get away with this because the system has a very fast response time, so that it can react quickly to any overshoot. Furthermore, a resistor and capacitor between the output and the 0 V rail provides the system with enough inertia to keep the oscillations small. So although the output hunts, the amplitude of the oscillation is too small to make any real difference.

Efficiency

The voltage regulator circuit of Fig 3.21 does not use energy very efficiently. Consider the voltage at C. If V_{th} for the MOSFET is 1 V, then C needs to sit at about 5 + 1 = 6 V in normal operation. But C can only get to within 2 V of its top supply rail, so the system can't operate if the voltage at the input is less than about 6 + 2 = 8 V. This is shown in the transfer characteristic of Fig 3.21. So current from the input has to drop through at least 3 V before it gets to the output terminal, giving a maximum energy efficiency of only about 60%.

Unstabilised supply

Voltage regulators are designed to run off **unstabilised supplies**, systems that run off the mains whose output is d.c. but not necessarily at a fixed voltage. Provided that the voltage of the unstabilised supply is high enough to power up the voltage regulator, it will be able to hold its output terminal at a steady fixed voltage, regardless of variations in the voltage of the unstabilised supply. The block diagram for a typical unstabilised supply is shown in Fig 3.22.

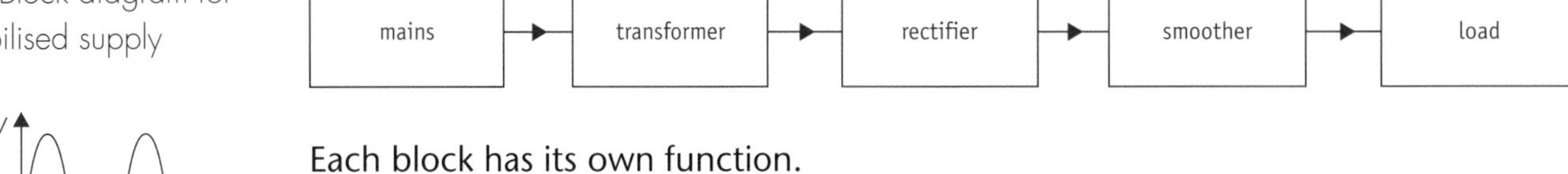

Fig 3.22 Block diagram for an unstabilised supply

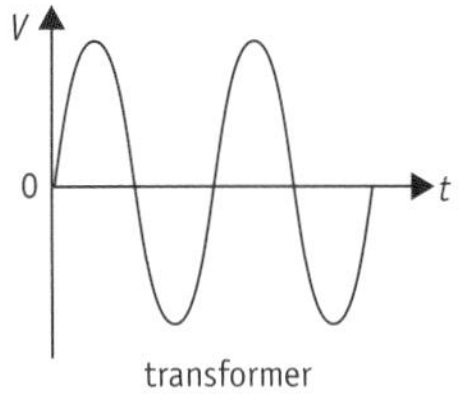

rectifier

smoother

Fig 3.23 Voltage–time graphs for signals at the output of three blocks

Each block has its own function.

- The **mains** provides an a.c. signal with an amplitude of 325 V at a frequency of 50 Hz.
- The **transformer** reduces the amplitude of the a.c. to about 12.5 V.
- The **rectifier** converts the a.c. into d.c., so that the current at its output flows only in one direction. The peak output is about 11 V.
- The **smoother** ensures that the voltage across the **load** never drops below a certain value.

The graphs of Fig 3.23 illustrate the effect of the rectifier and smoother on the signals which pass through them.

Supply circuit

A possible circuit for an unstabilised supply is shown in Fig 3.24.

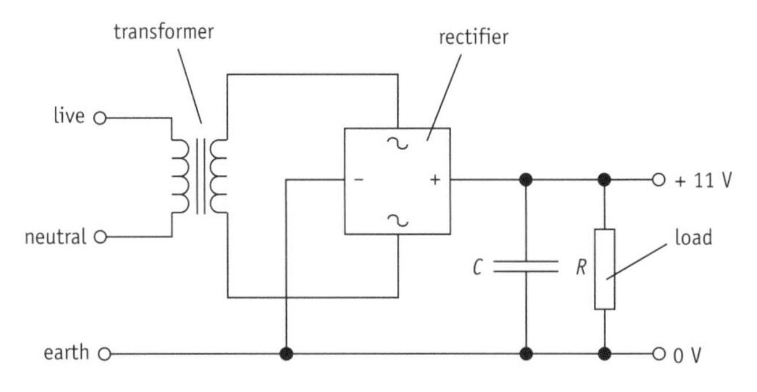

Fig 3.24 Circuit diagram for an unstabilised supply

The mains supply comes in from the left on the **live** and **neutral** lines. These carry the high voltage a.c. to the **primary coil** of the transformer. The **secondary coil** feeds a.c. at a much lower (and safer) voltage to the rectifier. Notice that there is no direct electrical connection between the primary and secondary coils, allowing the transformer to provide **isolation** between the dangerous mains input and the safe low voltage output of the supply. The rectifier is a four-terminal device, with two a.c. inputs (marked ~) and two d.c. outputs (marked + and −). Charge can flow in or out of the a.c. inputs, but only in one direction at the d.c. outputs. So the + terminal only sources current into the load and the − terminal only sinks current from the 0 V **earth** rail. A capacitor in parallel with the load acts as the smoother, charging up from the + terminal and discharging through the load.

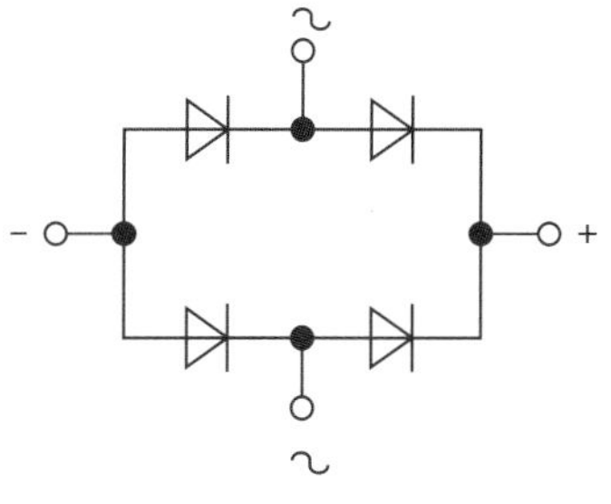

Fig 3.25 Inside a bridge rectifier

Rectifying and smoothing

The four diodes of a bridge rectifier (see Fig 3.25) operate in pairs to direct current from the transformer secondary to the top plate of the capacitor, regardless of the polarity of the a.c. Take a look at Fig 3.26 below. At the positive peaks of the a.c., charge flows through one pair of diodes to charge up the capacitor. At the negative peaks, charge flows through the other pair of diodes, also charging up the capacitor. At all other times the diodes are all reverse biased, so the capacitor has to discharge through the load. Of course, as the capacitor discharges, its voltage drops, so the time constant RC of the capacitor–load circuit needs to be chosen carefully, as follows.

The period of a 50 Hz mains supply is 20 ms.

$T = ?$

$f = 50$ Hz

$$f = \frac{1}{T}$$

$$T = \frac{1}{f} = \frac{1}{50} = 2.0\times10^{-2}\ \text{s or } 20\ \text{ms}$$

So there is a burst of current in the diode bridge every 10 ms as the capacitor charges. It therefore makes sense to set the value of RC to 100 ms, ten times as long as the discharge time. The value of R comes from the current to be drawn from the unstabilised supply rail at +11 V. Let's suppose that it is 0.5 A.

$R = ?$

$V = 11$ V

$I = 0.5$ A

$$R = \frac{V}{I} = \frac{11}{0.5} = 22\ \Omega$$

So when it draws 0.5 A, the load behaves like a 22 Ω resistor.

$C = ?$

$\tau = 100$ ms

$R = 22\ \Omega$

$$\tau = RC$$

$$C = \frac{\tau}{R} = \frac{100 \times 10^{-3}}{22} = 4.5\times10^{-3}\ \text{F or } 4\ 500\ \mu\text{F}$$

With this capacitor, the voltage of the output terminal will only fluctuate by 10%.

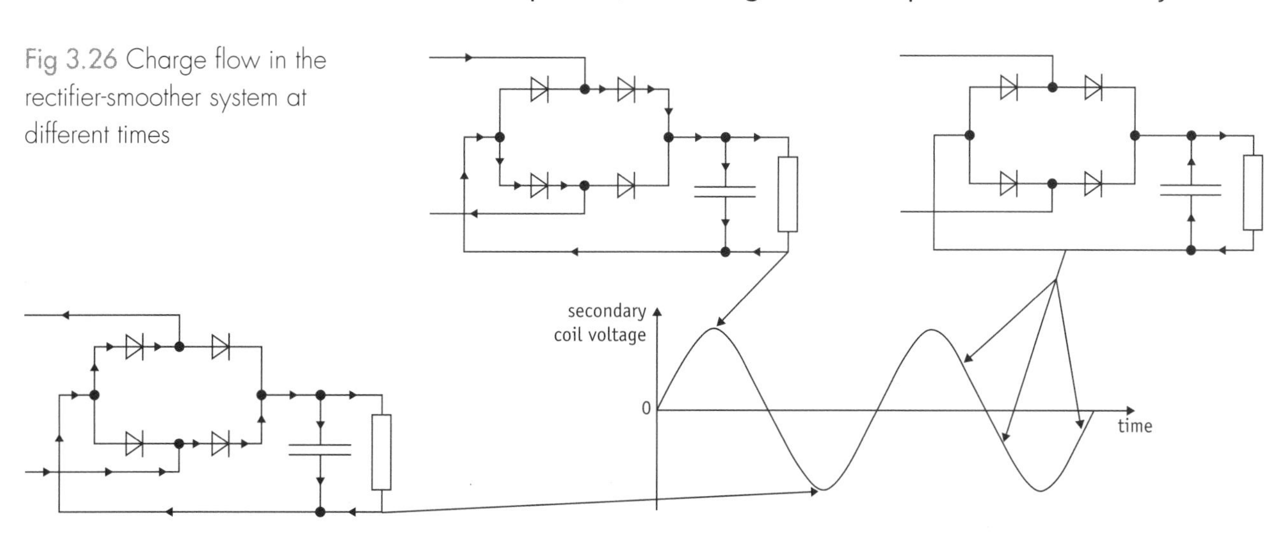

Fig 3.26 Charge flow in the rectifier-smoother system at different times

Switched-mode

The power supply described on the previous pages has one major disadvantage. The transformer is likely to be bulky and heavy. Increasing the frequency of the a.c. passing through it allows the size of a transformer to come down dramatically. So **switched-mode** power supplies increase the frequency of the mains a.c. from 50 Hz to 50 kHz before using a tiny transformer to reduce the voltage. The block diagram is shown in Fig 3.27.

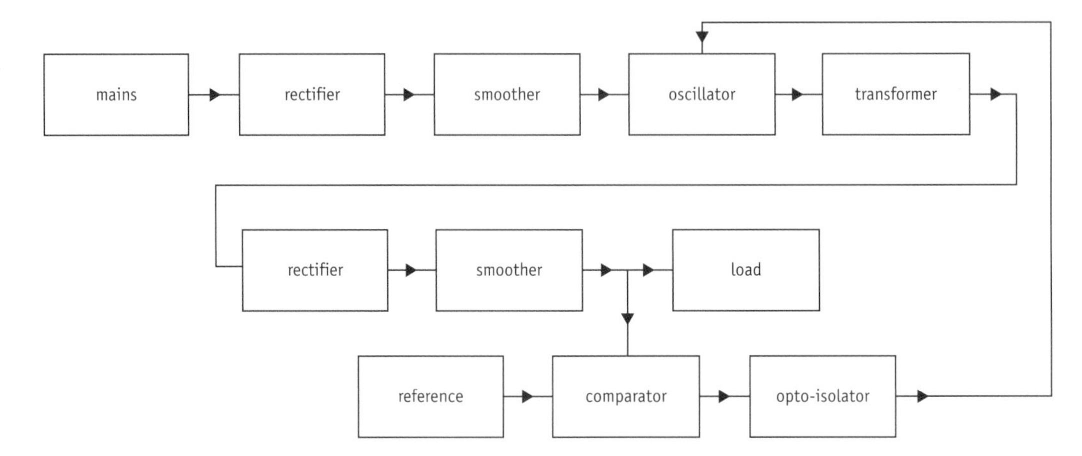

Fig 3.27 Block diagram of a switched-mode power supply

The diagram has three rows of blocks.

- The top row has the high voltage stuff. The mains is rectified and smoothed to provide the 50 kHz oscillator with supply rails 325 V apart. The high frequency a.c. is then reduced by the transformer.
- The middle row has the high power, low voltage stuff. The rectifier and smoother take the a.c. from the transformer secondary coil and use it to place a d.c. voltage across the load.
- The bottom row senses when the voltage across the load becomes too large and switches the oscillator off immediately. It does this via an **opto-isolator** to ensure that dangerous high voltage does not leak back into the low voltage circuitry.

There is another saving as well. The only heat losses occur as charge passes through the two rectifiers (virtually no energy loss occurs in the transformer), improving the energy efficiency of the system.

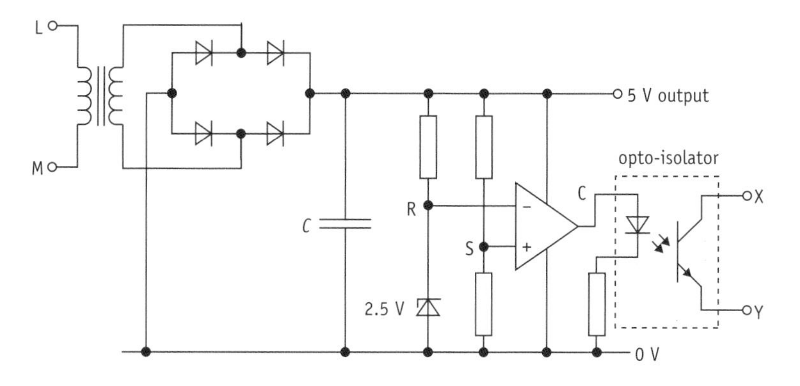

Fig 3.28 Low voltage part of a switched-mode power supply

Switch on, switch off

The low voltage end of the power supply comes to life as soon as the high voltage oscillator starts working. As you can see from Fig 3.28, the low voltage 50 kHz a.c. from the secondary coil of the transformer charges the capacitor *C* at intervals of 10 μs. As the voltage of the output line rises above 5 V, the voltage at S rises above the 2.5 V reference signal provided by the Zener diode. The op-amp output C immediately saturates high, sourcing current into the infrared LED of the opto-isolator. Infrared light passes from the LED to the phototransistor (there is no direct electrical connection between the two, even though they are in the same i.c. package), allowing it to sink current at X and source it at Y.

Now look at Fig 3.29, the high voltage end of the system. The oscillator is designed to turn off when X is connected to Y, so that it no longer feeds a 50 kHz a.c. signal of amplitude 325 V into the primary of the transformer via terminals L and M. Without any energy coming through the transformer, the voltage of the 5 V output line starts to drop as it sources current into any load attached to it. That drop will be sensed by the op-amp, allowing it to turn the oscillator back on via the opto-isolator. This on-off feedback places charge back onto the top plate of capacitor *C* as fast as it is lost to the output or the sensing circuitry. The short response time and relatively large inertia of the system usually results in very little overshoot.

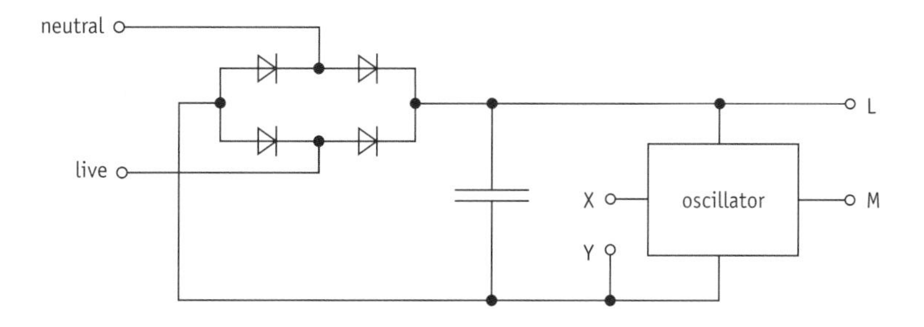

Fig 3.29 High voltage part of a switched-mode power supply

Questions

3.1 On-off control

1 A 12 V lamp provides light for the interior of a rat cage.

(a) Draw a block diagram for an open-loop system to control the light level inside the cage. Use some of the blocks from this list.

comparator power amplifier reference sensor switch transducer

(b) Draw a circuit diagram for the system. Describe its use to control the light level in the cage.

(c) Explain why a closed-loop system would provide better control of the light level.

(d) Draw a block diagram for a closed-loop system to control the light level. Choose more of the blocks from the list in (a).

(e) Draw a circuit diagram for the closed-loop system.

2 An on-off control system is used to control the temperature of a room.

(a) Draw a block diagram for the system.

(b) Describe the function of each block in the system.

(c) Draw a graph to show how the temperature of the room changes with time when the control system is operating. Explain the shape of the graph.

(d) Draw a circuit for the sub-system which senses the temperature of the room. Explain how it operates.

3 The block diagram of Fig 3.30 is for the speed sensor of an electric motor. The output of the magnetic sensor goes low each time a magnet on the motor shaft passes next to it.

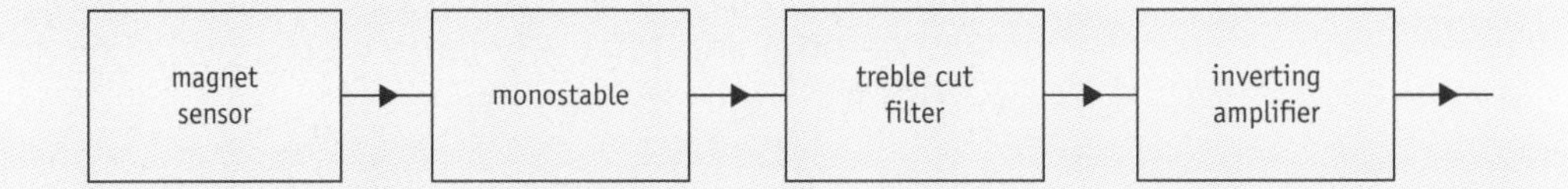

Fig 3.30 Block diagram for question 3

(a) The maximum speed of the shaft is expected to be 3000 r.p.m. Explain why a pulse length of 20 ms is a good choice for the monostable.

(b) The monostable produces a 20 ms high-going 5 V pulse each time it is triggered by the magnet sensor. Draw a circuit diagram for the monostable. Justify all component values.

(c) The filter is required to have a break frequency of 2 Hz. Draw a circuit diagram for it and justify all component values.

(d) Draw a circuit which uses this sensor to control the speed of the electric motor. Explain the operation of the circuit.

3.2 Proportional control

1 A d.c. motor is used to move an object along a line.

(a) Draw the circuit diagram of a system which uses proportional control to set the position of the object.

(b) Summarise the system as a block diagram. Explain the function of each block.

(c) An on-off control system could have been used to set the position of the object instead. Explain the advantage of using proportional control. Include position–time graphs for both systems.

2 This question is about the ramp generator of Fig 3.31.

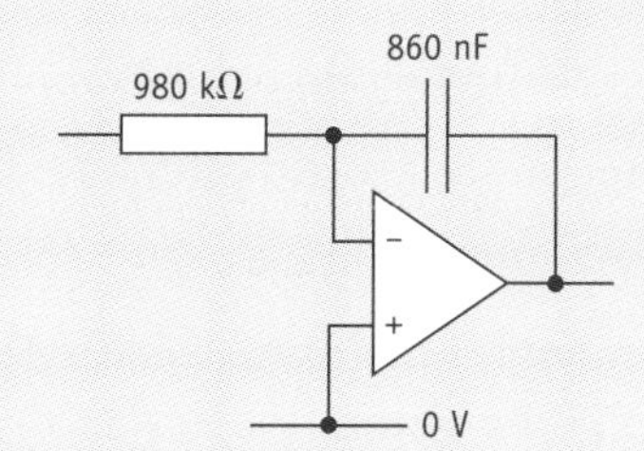

Fig 3.31 Circuit for question 2

(a) Show that the output will rise by 6 V/s when the input is held at −5 V.

(b) At a particular time, the voltage at the output is +4 V. If the input is held at +8 V, what is the voltage at the output after a time interval of 1.5 s?

(c) Explain why the output of the circuit does not change when the input is held at 0 V.

3 A proportional control system is used to maintain a constant light level in a rat cage, with a lamp as the light source.

(a) Draw a block diagram for the system.

(b) Draw a circuit diagram for the system. Show that it provides negative feedback.

(c) How is the system better than an on-off control system?

(d) When the door of the cage is opened, the light level suddenly rises. Use voltage–time graphs to show how the proportional control system responds to this change.

3.3 Power supplies

1 A d.c. voltage regulator produces a steady +6 V supply from an unstabilised supply of between +9 V and +15 V.

(a) Draw a block diagram for the system. Explain how the system operates.

(b) Draw a circuit diagram for the system.

(c) The Zener diode has a maximum power rating of 250 mW. Show that a series resistor of 470 Ω is a safe value to use.

(d) Explain why it is unnecessary for your system to use proportional control.

2 An unstabilised power supply uses a transformer, diode bridge and capacitor to convert an a.c. signal from an oscillator into a d.c. one across a load.

(a) Draw a block diagram for the system. Describe the function of each block.

(b) Draw a circuit diagram of the power supply.

(c) The a.c. output of the transformer has a peak voltage of 20 V and a frequency of 60 Hz. Sketch graphs, on the same axes, of the output of the transformer and the voltage across the load. Add scales to both axes.

(d) Explain the shape of the graph of the voltage across the load.

(e) If the load draws a maximum current of 600 mA, calculate a suitable value for the smoothing capacitor.

3 This question is about switched-mode power supplies.

(a) Draw the block diagram for a switched-mode power supply.

(b) Explain the function of each block.

(c) Draw the circuit symbol for the opto-isolator. Describe its transfer characteristic. Explain its function in the system.

(d) Draw the circuit symbol for the transformer. Describe its transfer characteristic. Explain its function in the system.

(e) The supply is required to maintain a steady +12 V at its output, using a Zener diode rated at 5 V, 250 mW to provide the reference signal. Draw a circuit diagram for the sensor, reference and comparator blocks. Give all component values and justify them.

Learning summary

By the end of this chapter you should be able to:

- understand the difference between open-loop and closed-loop control
- draw block diagrams for on-off and proportional control systems
- design sensors for light, temperature, speed and position
- use graphs to describe the response of on-off and proportional control systems
- understand the operation of a ramp generator and difference amplifier
- use a diode bridge and capacitor to generate an unstabilised power supply
- design a voltage regulator
- draw a block diagram for a switched-mode power supply and explain how it works

CHAPTER 4

Microcontroller systems

4.1 Hardware

You will recall from your work at AS, that the behaviour of a **microcontroller** (or **PIC**) is determined by the sequence of binary words stored in its memory. That sequence (the **program**) can be put together from a flowchart by a computer and downloaded into the microcontroller via a programming board. This chapter aims to further deepen your understanding of microcontrollers:

- How a microcontroller can be assembled from familiar sub-systems such as memories, registers and processors.
- How the stored program affects the operation of the system.
- How to write programs in assembler.

Although many different microcontrollers are available, once you have understood how to use one, then it is relatively easy to adapt to another.

Inputs and outputs

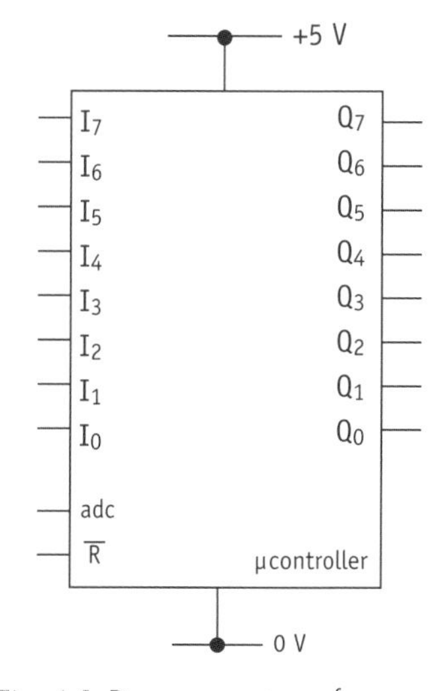

Fig 4.1 Pin connections for the microcontroller

Fig 4.1 shows the external connections to the microcontroller to be considered for the rest of this chapter. Many microcontrollers can be made to behave like the one featured in this chapter. Although they may all be put together with a different organisation of memory, ports and buses, the end product is the same.

There are five different connections:

- **Supply rails** are connected at +5 V and 0 V.
- An active-low **reset** pin is labelled $\overline{R}$. Pulsing this pin low forces the PIC to go to the start of its program.
- The **input port** has eight digital inputs labelled I_7 to I_0. Bytes placed at those pins can be read into the system as required by the program.
- The **output port** has eight digital outputs labelled Q_7 to Q_0. The program can write bytes to these pins, where they are latched.
- An **analogue input** is labelled adc. This can be connected to an **analogue-to-digital converter** inside the microcontroller which codes any voltage between 0 V and 5 V at the analogue input as an eight-bit word.

Ports and buses

The first step in our top-down analysis of a microcontroller is shown in Fig 4.2.

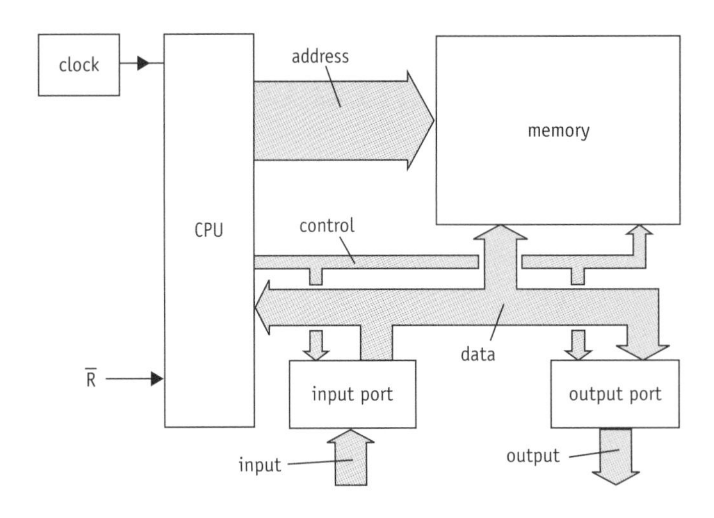

Fig 4.2 Block diagram for a microcontroller

As you can see, it consists of five sub-systems connected to each other by three buses. Let's start with the most complicated one.

- The **central processing unit** (or **CPU**) lies at the centre of the system. It is the only block which can place words on the **control** and **address** buses, so it dictates the behaviour of the other sub-systems.
- The **clock** feeds a constant train of pulses into the CPU at a frequency of 8 MHz. The CP U uses this signal to synchronise all of its operations.
- The address bus has 10 lines, so the **memory** has $2^{10} = 1024$ locations, each of which holds a byte, an eight-bit word. The memory is split into two halves, as shown in the memory map of Fig 4.3. The bottom half, from addresses 000 to 1FF is flash memory. This holds the program. The top half, from 200 to 3FF, is SRAM.
- The **input port** is effectively a bank of eight tristates which can be enabled by the CPU via the control bus, allowing bytes at the input pins to get to the data bus.
- The **output port** is effectively a bank of eight flip-flops which can latch a byte from the data bus and hold it at the output pins.

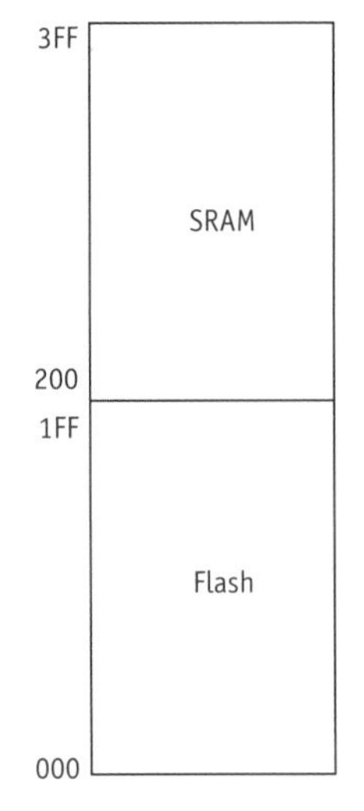

Fig 4.3 Memory map

Before you find out how these subsystems interact with each other to execute the program stored in the memory, you need to find out more about the interior of the CPU.

Registers

As its name implies, the CPU is where binary words are combined by logic gates to make other binary words. Those words have to be stored in **registers** before and after these operations. The action of the **processor** is also decided by a word which also has to be stored in a register. Fig 4.4 on the next page shows that there are three distinct classes of register.

Fig 4.4 Registers in the CPU

program counter PC
stack pointer SP
memory pointer MP
instruction register IH
instruction register IL

general purpose S7
general purpose S6
general purpose S5
general purpose S4
general purpose S3
general purpose S2
general purpose S1
general purpose S0

- The **program counter**, **stack pointer** and **memory pointer** each hold addresses for the memory, so need to be ten bits wide.
- The two **instruction registers** are each eight bits wide. Their contents determine the process to be carried out by the CPU on the next clock pulse.
- Eight **general purpose** registers hold bytes for processing by the CPU.

The contents of all these registers are cleared to zero each time that the reset line $\overline{R}$ is pulled low.

ALU

All the processing of words in the CPU is done by the **arithmetic logic unit** (or **ALU**). As you can see from Fig 4.5, the ALU can take in a byte from any one of the general purpose registers **S7** to **S0**. The contents of the **IH** register determines which registers are accessed and the operation to be performed on them when the ALU is activated. Here are a few of the many operations possible.

- Copy the contents of IL into S7.
- Add S0 to S3 and store the result in S0.
- Increment S4.
- Shift the contents of S2 one place to the right.
- Copy the contents of S5 into S6.
- Perform the AND operation on S1 and S4, storing the result in S1.
- Copy the input port to S7.
- Copy S2 to the output port.

For simplicity, none of the control lines have been shown in Fig 4.5.

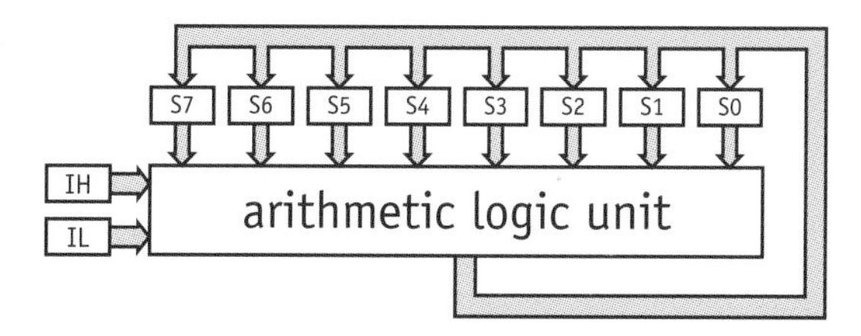

Fig 4.5 The ALU processes the words in the general purpose registers

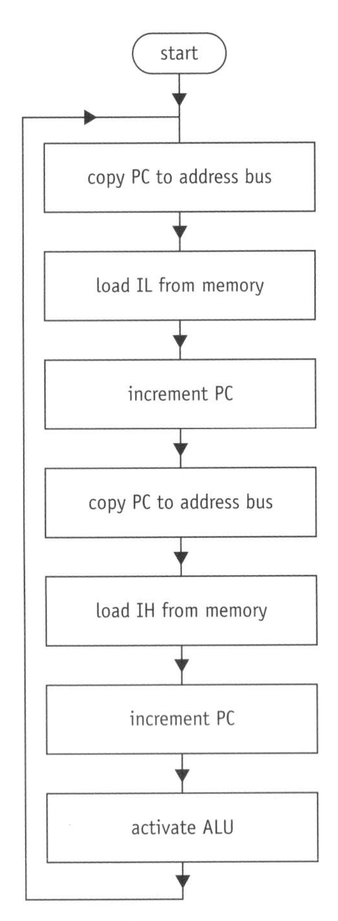

Fig 4.6 Machine cycle flowchart

Reset and run

So what happens immediately after the microcontroller has been reset? It goes straight into a short routine, known as a **machine cycle**, summarised in the flowchart of Fig 4.6. It has only eight steps.

1. Copy the contents of the program counter (or **PC**) to the address bus. First time round, this will access the memory location whose address is 000.
2. Use the control and data buses to read the byte stored in location 000 and store it in the instruction register **IL**.
3. Add one to the contents of PC. This **increment** operation ensures that the program is read out of the memory one byte after another, from the bottom up.
4. Copy the contents of PC to the address bus. This will access location 001.
5. Use the control and data buses to copy a byte from memory into **IH**.
6. Increment PC once more.
7. Active the ALU.
8. Return to step 1.

And that's all the CPU ever does – just goes round this routine a million times a second. It only stops when $\overline{R}$ goes low or the power supply is turned off. Notice how the contents of the program counter have increased by two at the end of each cycle, so each cycle loads a different pair of bytes from the memory into the instruction registers. It is those pairs of bytes which make up the program stored in the flash memory.

Program codes

The table shows the first few bytes of a program stored in memory. Each column shows something different.

- The first column is the **address**, in hexadecimal, of each byte in memory. Notice that it starts from 000, and goes up by 001 each time.
- The second column contains the program **code**, again in hexadecimal. Codes at odd addresses get loaded into IH, so represent commands for the ALU. Codes at even addresses get loaded into IL, so act as data for the ALU.
- The third column contains **assembler codes**. Appropriate software (called an **assembler**), run on a host computer, can translate these into program codes.
- The fourth column describes the effect of each step of the program on the system.

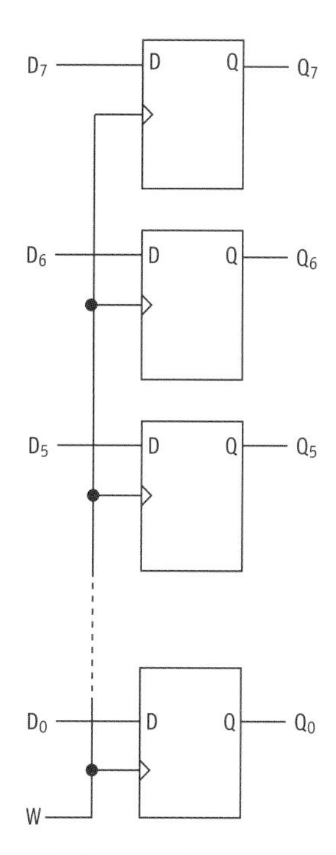

Fig 4.7 Flip-flops making an output port

address	code	assembler	function
000	00	MOVI s0,$00	Place the byte 00 into register s0.
001	E0		
002	04	OUT $04,s0	Copy s0 into the output register at address 04. This sets up Port B to act as an input port.
003	B9		
004	FF	MOVI s0, $FF	Place the byte FF into register s0.
005	E0		
006	0A	OUT $0A,s0	Copy s0 into the output register at address 0A. This sets up Port D to act as an output port.
007	B9		
008			
009			

Although assembler codes look quite unfriendly, you will get used to them. After all, it is easier to remember MOVI, rather than E0, as the code that means move a byte into a register!

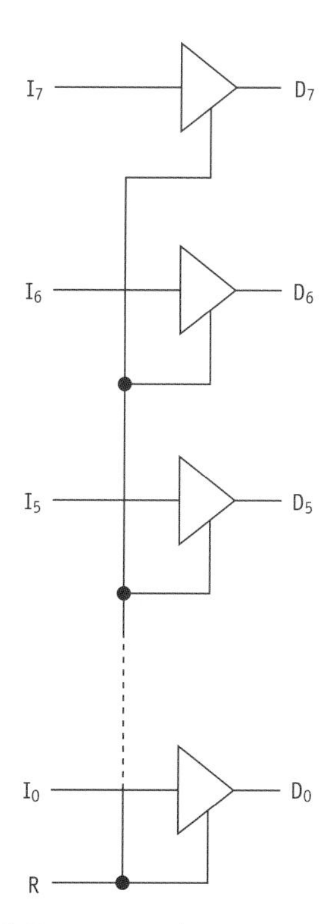

Fig 4.8 Tristates making an input port

Ports

The series of codes above is an example of a **program header**. This is a sequence of codes which needs to go at the start of every program, setting up the micro-controller so that it is ready to execute the main program which starts at address 008. It will be different for each type of microcontroller. This particular header sets up two of the ports of the microcontroller to act as input and output. It is standard practice for a microcontroller to have at least two ports, each of which can act as input or output, depending on the contents of the appropriate **output registers** in the CPU. Take a look at the arrays of flip-flops and tristates shown in Figs 4.7 and 4.8. Although each port contains both of these in parallel between the data bus and the i.c. pins, only one is activated at any one time. So if the port is set as an output, then a rising edge at W latches the contents of the data bus D_7 to D_0 onto the output pins Q_7 to Q_0. Conversely, if the port is set as an input, then each time R goes high the tristates connect the input pins I_7 to I_0 to the data bus.

Assembler program

Here is an example of a short program written in assembler. All bytes are in hexadecimal.

```
        .equ   Q = 0B       ;Q is the output port label
        .equ   I = 03       ;I is the input port label
        .org   000          ;start program at 000
        movi   s0, 00
        out    04, s0       ;set up port B as input
        movi   s0, FF
        out    0A, s0       ;set up port D as output
start:  in     s0, I        ;copy input port to s0
        out    Q, s0        ;copy s0 to output port
        jp     start        ;repeat main program
```

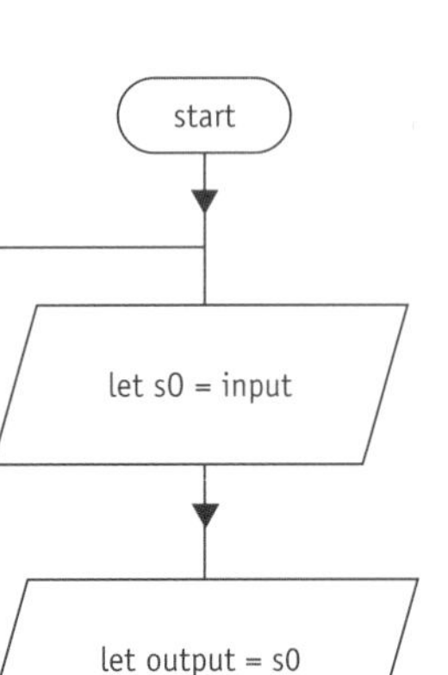

Fig 4.9 Flowchart for the assembler program

The first three lines contain commands for the assembler program itself. The two .equ commands mean that Q and I will be read as 0B and 03 respectively, saving you the bother of remembering that these are the addresses of the output and input ports. The .org command instructs the assembler to place the start of the program at the bottom of memory.

The next four lines contain the header to set up the input and output ports. Notice the use of semicolons to add useful comments.

The next line starts with a **label**. You can tell that it is a label because it ends with a colon. This particular label has been chosen to remind you that this is where the main program starts. The flowchart of Fig 4.9 summarises the effect of running that program – continually passing a byte from the input port to the output port.

Hex code

The job of an assembler is to take a program written with assembler codes and translate it into a **hex file**, ready for uploading into the flash memory of the microcontroller. This process is known as **assembly**, and leads to this **listing** for the main program.

address	hex code	assembler
008	B1 03	in s0, I
00A	B9 0B	out Q, s0
00C	CF FA	jp start

Notice how the codes have been grouped in pairs. The first byte of each pair will be loaded into IH, determining the action taken by the ALU. The second byte loaded into IL only provides data. So B1 makes the ALU copy the contents of a port into s0, and 03 tells it which port to use. Similarly, B9 makes the ALU copy s0 to a port and 0B selects the port accordingly.

Jumps

You may have noticed that the assembler program on the previous page contains only one address, in the third line. That is because, as it assembles the program into a hex file, the program sorts out the addresses automatically. So the point labelled 'start' ends up placed at address 008 in the listing above. So how does the assembler sort out a jump back to that point in the last line of the program? Consider the left-hand memory map shown in Fig 4.10. It shows the situation just **after** both bytes of the jump instruction CF FA have been copied to the instruction registers. The program counter contains 00E, as shown by the arrow **pointing** to that address.

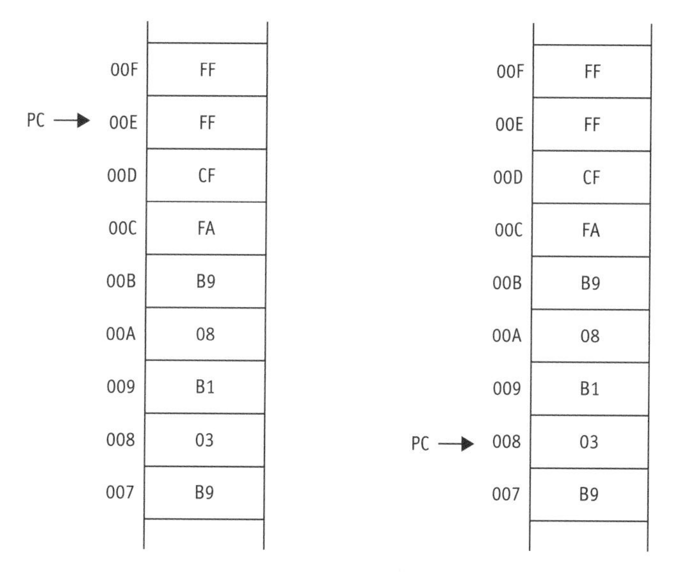

Fig 4.10 A jump instruction changing the program counter

The jump instruction needs to change the contents of PC so that it points to address 008, as shown in the right-hand memory map. This is six locations further down the memory (count them). So the jump instruction needs to add −6 to the current contents of the program counter. So what is −6 in hexadecimal? Here are the steps involved in the conversion.

- Write down +6 as an eight bit word in BCD: 0000 0110.
- Invert all the bits: 1111 1001
- Add one: 1111 1010
- Convert to hexadecimal: FA

So the assembler arranges for IH to be loaded with CF (add the contents of IL to PC, storing the result in PC) and IL with FA.

4.2 Software

By now you should appreciate how a microcontroller can be assembled from simpler subsystems, such as memory, registers, tristates and logic gates. This section of the chapter is going to show some programming techniques which will allow you to make full use of the device's potential. You can find a full list, in Appendix E, of the few assembler codes used.

Test and skip

Suppose the following program is loaded into the microcontroller of Fig 4.11. What happens after a reset?

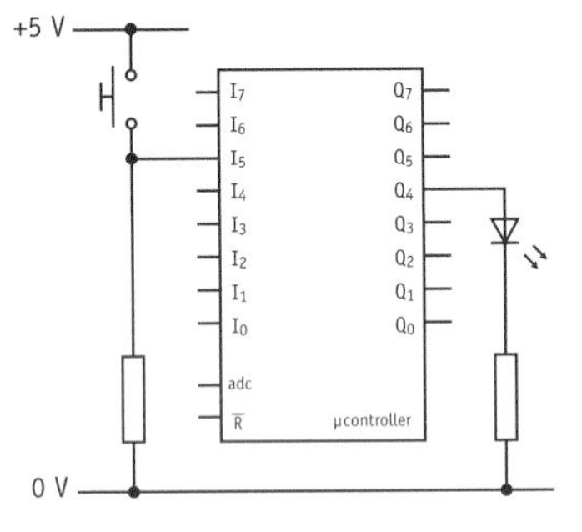

Fig 4.11 How can I_5 be tested separately from the other input pins?

```
start:    movi  s4, 20     ;load s4 with 0010 0000
          movi  s7, 10     ;load s7 with 0001 0000
          movi  s6, 00     ;load s6 with 0000 0000
back1:    out   Q, s6      ;turn off LED
back2:    in    s0, I      ;copy input port to s0
          and   s0, s4     ;mask for pin 5
          jz    back1      ;jump if pin 5 low
          out   Q, s7      ;pin 5 high, so turn on LED
          jp    back2      ;go back and test pin 5 again
```

The first three lines use the MOVI (short for **move-immediate**) command to place a series of bytes in three registers. One of them (s6) is then copied to the output port, making all eight pins low. The LED does not glow.

Now the input port is copied to s0. Pin 5 is connected to a pull-down resistor and a switch, so it is definitely high (when the switch is pressed) or low. But what of the others? We don't know, so the next command uses the AND operation to **mask** off all the bits of s0 except for bit 5. Each bit of s0 is combined with s4 by the AND operation, storing the result back in s0. Remember, 0.X = 0 but 1.X = X.

$$I_7I_6I_5I_4\ I_3I_2I_1I_0 \text{ AND } 0010\ 0000 = 00I_50\ 0000$$

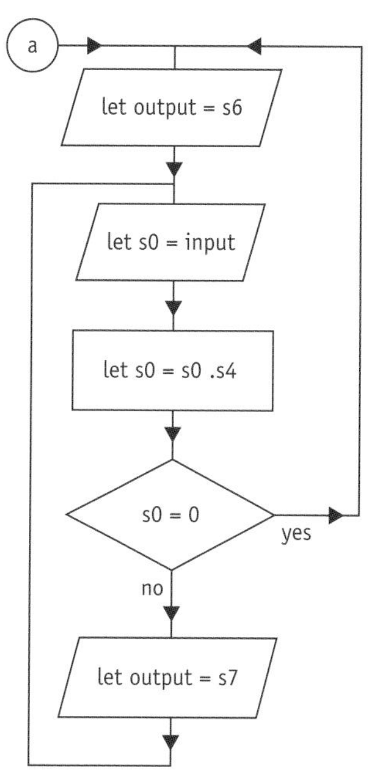

Fig 4.12 Flowchart for the test-and-skip loop

Conditional jump

The next command **jz** (or **jump if zero**) is a **conditional jump**. It will only change the program counter if the last operation results in zero. Because of the masking operation, the only bit which can be 1 is I_5, and this only happens when the switch is pressed. So, provided that the switch is not being pressed, the program loops to the label 'back1' – and gets stuck. The system gets trapped in the right-hand loop of the flowchart shown in Fig 4.12.

Now suppose that the switch is pressed. I_5 goes high, so that one of the bits of s0 remains high after the AND operation. It fails the condition for the jump to 'back1', leaving the program counter unaltered, allowing s7 to be copied to the output port. This pushes Q_4 high, turning on the LED. Finally, the program makes the system have another look at the input port – and ends up trapped in the left-hand loop of the flowchart until the switch is released.

Toggling

Suppose that this program is loaded into the microcontroller of Fig 4.11 instead. What happens?

```
start:   movi  s4, 20    ;load s4 with 0010 0000
         movi  s7, 10    ;load s7 with 0001 0000
         movi  s6, 00    ;load s6 with 0000 0000
back1:   out   Q, s6     ;turn off LED
back2:   in    s0, I     ;copy input port to s0
         and   s0, s4    ;mask for pin 5
         jz    back1     ;jump if pin 5 low
         eor   s6, s7    ;change the state of bit 4
         out   Q, s6     ;copy current byte in s6 to output port
         jp    back2     ;go back and test pin 5 again
```

Notice the extra line that has been placed after the test we used to see if s0 was zero after the AND operation. If s0 turns out not be zero, the s6 register is subjected to the EOR operation with the contents of s7 (0001 0000). This **toggles** bit 4 of s6, changing its state. This results in the LED being turned continually on and off whenever the switch is being pressed. Opening the switch leaves the LED on or off at random, depending on when the switch was released.

Counting edges

The system shown in Fig 4.13 counts the number of times that the switch is pressed, displaying the result in binary on a set of eight LEDs.

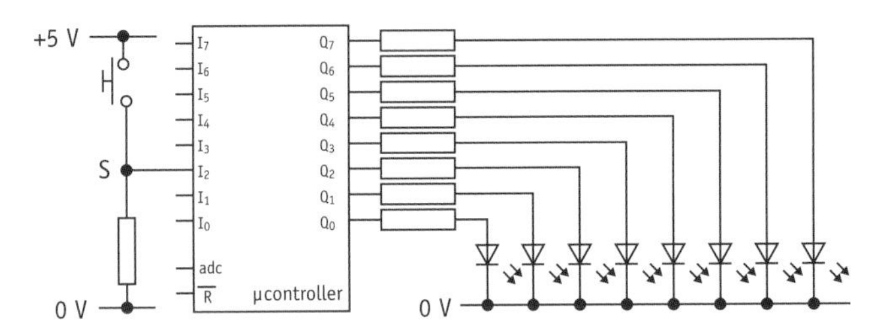

Fig 4.13 The rising edges at S are counted and displayed in binary

Here is the program. The first chunk sets up data in registers and turns all of the LEDs off.

```
start:    movi  s2, 04     ;set up mask of 0000 0100 for input test
          movi  s7, 00     ;zero s7 which will hold the count
back5:    out   Q, s7      ;output current state of the count
```

Now follows a test-and-skip loop to establish that S is high.

```
back1:    in    s0, I      ;start testing for S high
          and   s0, s2     ;s0 contains 0000 0S00
          jz    back1      ;loop back until S is high
```

S goes high because the switch is pressed. Switches bounce, so S will be unstable for a short time. This next chunk makes the system waste time by counting down from 250 to 0. Note the use of the **jnz** (or **jump if not zero**) command.

```
          movi  s6, FA     ;set up delay factor of 15 × 16 + 10 = 250 in s6
back2:    dec   s6         ;decrement s6
          jnz   back2      ;until it reaches zero
```

The system then looks to see if the switch has been released.

```
back3:    in    s0, I      ;start testing for S low
          and   s0, s2     ;s0 contains 0000 0S00
          jnz   back3      ;loop back until S is low
```

Another **delay loop** deals with instability at S when it is released.

```
          movi  s6, FA     ;set up delay factor in s6
back4:    dec   s6         ;decrement s6
          jnz   back4      ;until it reaches zero
```

The final chunk updates the count in s7 and jumps back

```
          inc   s7         ;increase value of count by one
          jp    back5      ;jump back to update count display
```

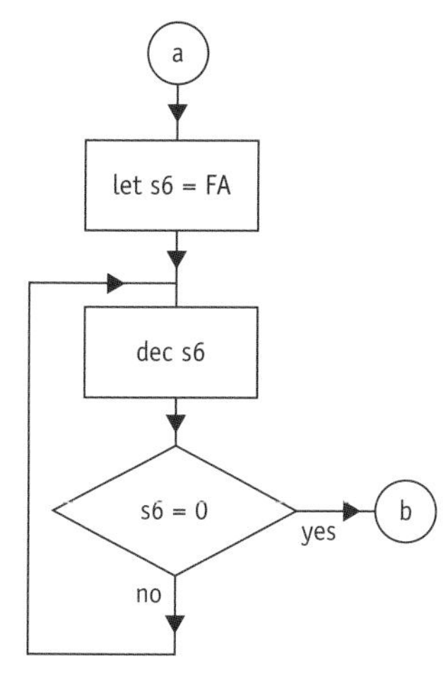

Fig 4.14 Flowchart for the time delay chunk

Time delay

How much time does the system waste in each delay loop? One machine cycle is needed to execute each instruction, and an 8 MHz clock speed results in a time of 1 μs for each machine cycle. The loop contains two instructions (dec and jnz), and the system has to go around the loop 250 times, giving 500 instructions altogether. The whole chunk of Fig 4.14 makes the system waste time for about 0.5 ms.

Wasted space

The time delay routine is used twice in the program. This is wasteful, needlessly using up valuable limited space in flash memory. So it would be better to write the time delay routine as a **subroutine**, somewhere after the end of the main program, starting with a suitable label, as shown below. Then every time you want to use the time delay, you just use the **rcall** (or **routine call**) command followed by the appropriate label. The **ret** (or **return**) command at the end of the routine ensures that the system goes back to the main program once the delay routine has finished.

```
start:   movi   s2, 04     ;set up mask of 0000 0100 for input test
         movi   s7, 00     ;zero s7 which will hold the count
back5:   out    Q, s7      ;output current state of the count
back1:   in     s0, I      ;start testing for S high
         and    s0, s2     ;s0 contains 0000 0S00
         jz     back1      ;loop back until S is high
         rcall  delay      ;call time delay subroutine
back3:   in     s0, I      ;start testing for S low
         and    s0, s2     ;s0 contains 0000 0S00
         jnz    back3      ;loop back until S is low
         rcall  delay      ;call time delay subroutine
         inc    s7         ;increase value of count by one
         jp     back5      ;update output display

delay:   movi   s6, FA     ;set up delay factor in s6
loop4:   dec    s6         ;decrement s6
         jnz    loop4      ;until it reaches zero
         ret               ;return to main program
```

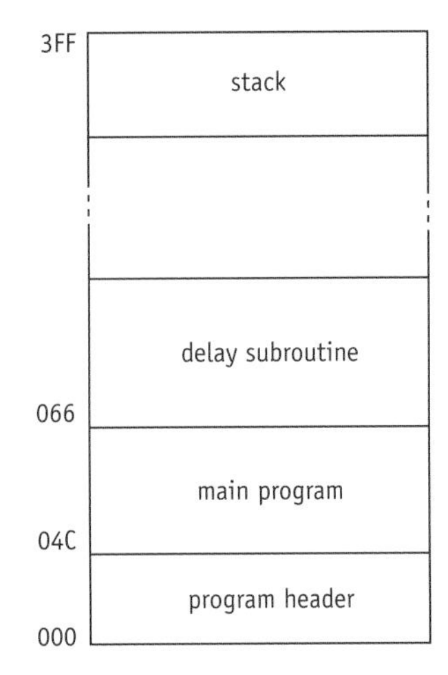

Fig 4.15 The stack is at the top of SRAM

Stack, push and pop

Each time that a subroutine is called, the system has to store an address in the **stack.** This is a block of memory at the top of SRAM. Without this, the system would never be able to get back to where it left off in the main program. So before any subroutines are called, the **stack pointer** (or **SP**) must be set to point to the top of SRAM (03FF). You only need to do this once, so it can safely be done as part of the program header. Here is an example.

```
movi   s0, 03       ;high byte of SP
out    3E, s0
movi   s0, FF       ;low byte of SP
out    3D, s0
```

The memory map of Fig 4.15 shows a main program at address 04C, followed by a subroutine at address 066. This is what happens when the delay subroutine is called with the program counter pointing to address 058.

- The low byte of PC (58) is stored at 3FF.
- SP is decremented to 3FE.
- The high byte of PC (00) in is stored at 3FE.
- SP is decremented to 3FD.
- The word 0066 is placed in PC.

The above sequence **pushes** the current contents of the program counter onto the stack and replaces it with 066, the address of the first instruction of the subroutine. At the end of the subroutine, the ret command **pops** the return address off the stack, as follows.

- Increment SP to 3FE.
- Copy 00 from the stack into the high byte of PC.
- Increment SP to 3FF.
- Copy 58 from the stack into the low byte of PC.

Notice that this procedure leaves SP pointing to the top of SRAM once more, awaiting the next call of a subroutine. The memory maps of Fig 4.16 show what happens to SP before, during and after a subroutine call. Notice how the stack is used on a **last-in first-out** basis; the bytes have to be pushed and popped in different orders.

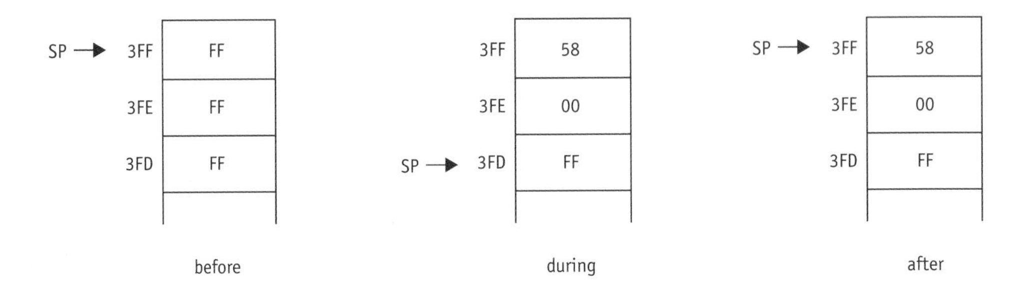

Fig 4.16 SP moves up and down the stack when a subroutine is called

Standard routines

The subroutine shown below provides a useful time delay of 1 ms, using the technique of **nested loops**.

```
wait1ms:   movi  s7, 02     ;initialise loop counter
loop2:     movi  s6, FA     ;start 500 µs delay
loop1:     dec   s6
           jnz   loop1
           dec   s7         ;decrement loop counter
           jnz   loop2      ;repeat delay until loop counter is zero
           ret              ;return to main program
```

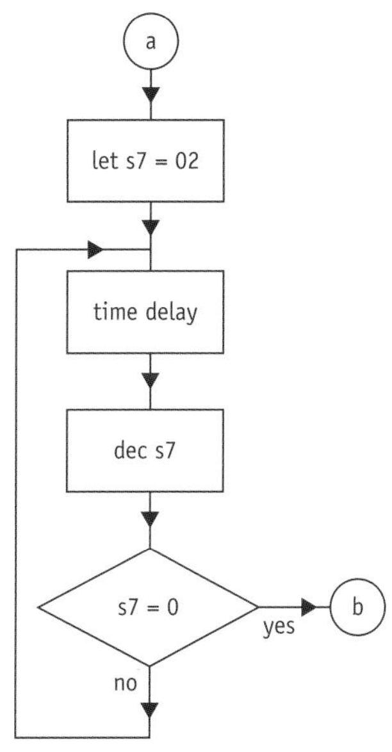

Fig 4.17 Flowchart for the wait1ms subroutine

The flowchart of Fig 4.17 shows how it works. Register s7 is set up as a **loop counter**, allowing it to keep track of many times the system goes around the central 500 μs delay loop. When using this subroutine you need to be aware of two things.

- It uses the s7 and s6 registers, so they shouldn't be holding any useful data for the main program when the subroutine is called. Both registers hold 00 when control returns to the main program.
- Each label must be unique, so it pays to have different styles of label for the main program and subroutines.

Information transfer

This subroutine counts the number of pins on the input port which are high, returning to the main program with the answer in register s5.

```
pincount:  movi  s6, 80     ;put 1000 0000 into the s6 mask
           movi  s5, 00     ;reset s5 counter to zero
loop3:     in    s7, I      ;copy input port to s7
           and   s7, s6     ;mask s7 with s6
           jz    loop4      ;jump ahead if bit is zero
           inc   s5         ;increment counter
loop4:     shr   s6         ;shift s6 one place to the right
           jnz   loop3      ;loop back to test next bit
           ret              ;return with count in register s5
```

The subroutine masks the byte at the input port eight times, using the **shr** (or **shift right**) command to move the 1 in the mask one place to the right each time. So s6 contains 80, 40, 20, 10, 08, 04, 02 and 01 in successive passes through the loop starting at loop3. The last shift operation leaves 00 in s6, allowing the system to return to the main program with the answer in s5.

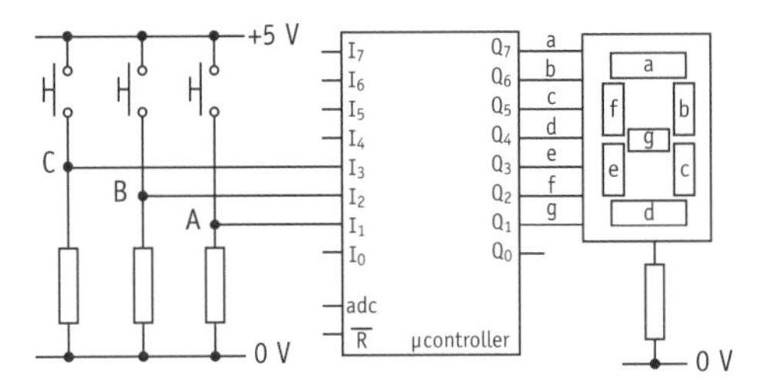

Fig 4.18 The display shows the number of switches being pressed

Look-up tables

The circuit of Fig 4.18 requires the seven segment LED to display, in decimal, the number of switches that are being pressed. Here is the program.

```
table:     FC, 60, 60, DA, 60, DA, DA, F2

start:     movi   s1, OE        ;prepare mask of 0000 1110 in s1
back1:     in     s7, I         ;copy input port to s7
           and    s7, s1        ;mask so that s7 holds 0000 CBA0
           shr    s7            ;right shift so that s7 holds 0000 0CBA
           rcall  readtable     ;call subroutine
           out    Q, s0         ;output byte in s0 returned by subroutine
           jp     back1
```

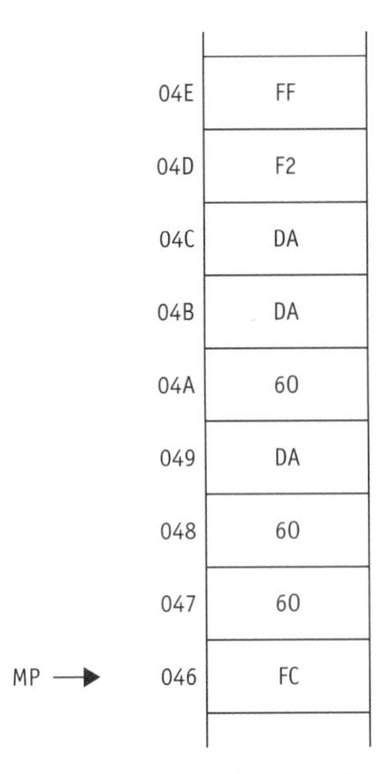

Fig 4.19 The look-up table

The assembler treats the bytes listed after the label **table:** as data, placing them in memory, starting at a particular location, as shown in Fig 4.19. The flowchart of Fig 4.20 shows how the **readtable** subroutine goes about retrieving a byte from this **look up table**. There are three separate steps.

- The **memory pointer** (or **MP**) points to the bottom of the look-up table.
- The number in s7 is added to MP, so that it points to a byte in the table.
- That byte is copied into s0, ready for transfer back to the main program.

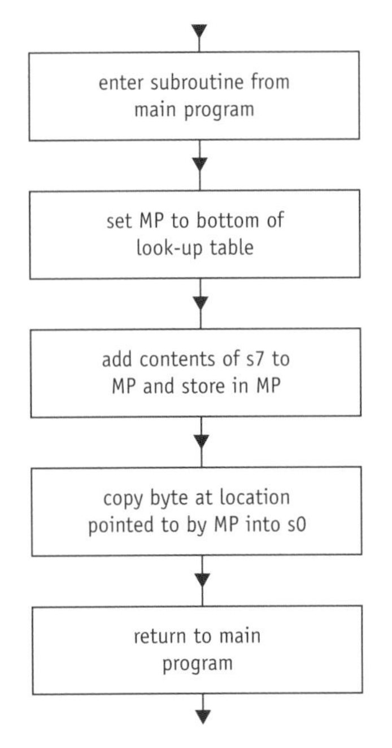

Fig 4.20 Flowchart for the readtable subroutine

For example, suppose that CBA is 110, so that two switches are pressed. Register s7 goes into the subroutine with the number six, moving the memory pointer six places up the look-up table, so that DA is copied into s0. As you can see from the table below, when DA is copied to the output port the seven-segment LED displays the number 2.

CBA	display	binary	hexadecimal
000	0	1111 1100	FC
001	1	0110 0000	60
010	1	0110 0000	60
011	2	1101 1010	DA
100	1	0110 0000	60
101	2	1101 1010	DA
110	2	1101 1010	DA
111	3	1111 0010	F2

Range detection

The circuit of Fig 4.21 is required to display H (for hot), C (for cold) or F (for fine), depending on the temperature of the thermistor. How do you set about writing the program?

Fig 4.21

The program will use the **readadc** subroutine to convert the voltage at the adc input into an eight bit word in s0, with 5 V giving FF and 0 V giving 00. The table shows the words required at the output port for each range of values in s0.

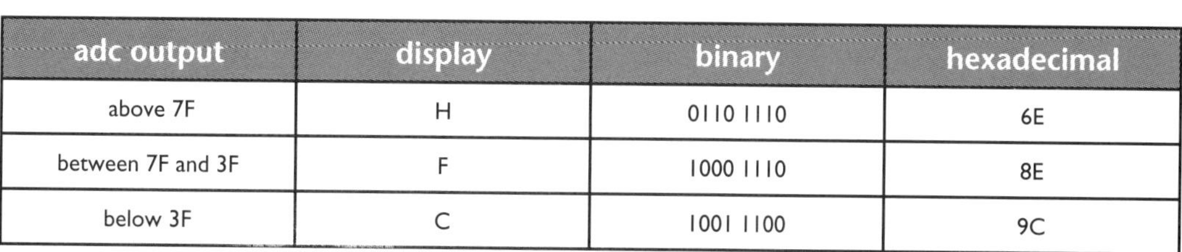

adc output	display	binary	hexadecimal
above 7F	H	0110 1110	6E
between 7F and 3F	F	1000 1110	8E
below 3F	C	1001 1100	9C

The first part of the program sets up the look-up table, reads in a byte from the readadc subroutine, and sets up the s7 for the readtable subroutine.

```
table:      6E, 8E, 9C
start:      rcall   readadc     ;byte in s0 codes for voltage at adc input
            movi    s7, 00      ;s7 will carry address of byte in look-up table
```

Now apply the first test by subtracting 7F from s0. If s0 was less than 7F, the result will be negative, leaving the most significant bit high. The **testmsb** subroutine increases s7 by one if this is the case.

```
            movi    s2, 7F
            sub     s0, s2      ;subtract 7F from byte
            rcall   testmsb     ;increment s7 if the result is negative
```

Now apply a second test to see if s0 was less than 3F.

```
            movi    s2, 3F
            rcall   readadc
            sub     s0, s2      ;subtract 3F from byte
            rcall   testmsb     ;increment s7 if the result is negative
```

Finally, use the readtable subroutine to display one of three bytes.

```
            rcall   readtable
            out     Q, s0
            jp      start
testmsb:    movi    s5, 80      ;prepare mask of 1000 0000 in s5
            and     s0, s5      ;s0 is 80 if number in s0 was negative
            jz      loop5       ;return if number in s0 was positive
            inc     s7          ;move one step up look-up table
loop5:      ret
```

Questions

4.1 Hardware

1 A microcontroller is made of a number of sub-systems which communicate with each other through buses.

(a) Explain what a bus is.

(b) Describe the functions of the address, data and control buses.

(c) Draw a block diagram to show how the CPU, memory, input port and output port are connected to each other by the address, control and data buses.

(d) Show how an input port can be made from tristates. Explain how it operates.

(e) Show how an output port can be made from flip-flops. Explain how it operates.

2 The CPU of a microcontroller contains a number of registers.

(a) Describe the function of a general purpose register.

(b) Show how a register can be constructed from flip-flops and tristates. Explain how to write words to it and read words from it.

(c) Describe the use of, and changes to, the program counter during a machine cycle.

(d) Explain why pulsing the reset line of a microcontroller low restarts its program.

(e) The clock feeds pulses to the CPU. What is the function of these pulses?

3 A particular microcontroller has the following specification for its memory.

- 2048 bytes of flash memory
- 512 bytes of SRAM

(a) How many lines will the address bus need?

(b) Explain why the address 000 needs to be flash memory.

(c) Draw a memory map for the system, showing the start and end addresses for each type of memory.

(d) Explain how a byte stored in memory can affect the behaviour of a microcontroller during a machine cycle.

(e) The program of a microcontroller is a sequence of bytes in its memory. Outline how that sequence can be generated and placed in memory.

4 The flowchart of Fig 4.22 features a conditional jump instruction.

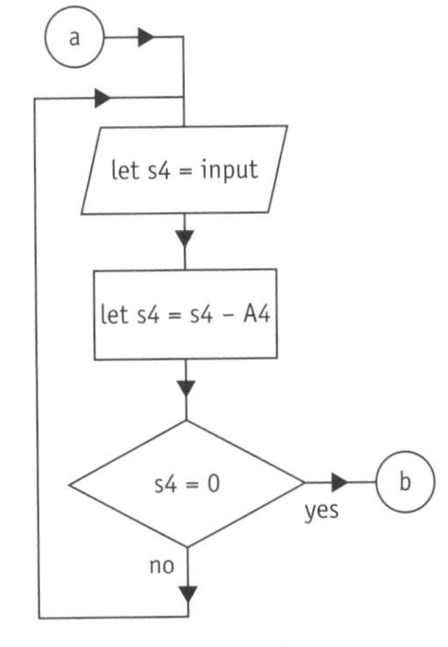

Fig 4.22 Flowchart for question 4

(a) Describe and explain the changes in the program counter when the jump happens.

(b) Use the instruction set of the Appendix to write the flowchart in assembler.

(c) Your assembler program contains a label. What happens to the label when the program is assembled into hex code?

Software

These questions use the set of assembler codes given in the Appendix.

1 This question is about the microcontroller arrangement of Fig 4.23.

Fig 4.23 Circuit for question 1

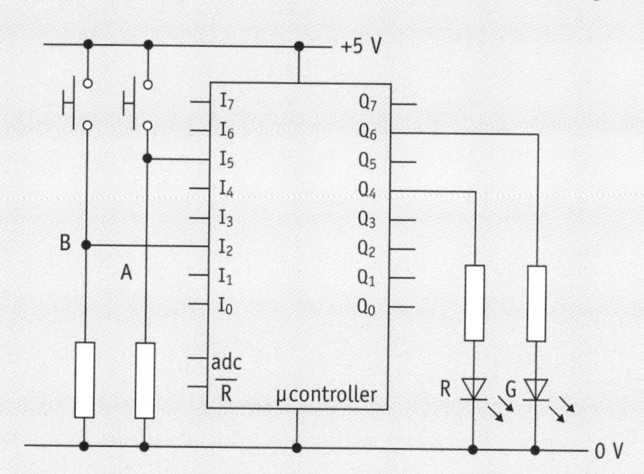

(a) Explain why the AND operation should be used to process any byte read in from the input port.

(b) Write an assembler program to make the LED labelled G glow whenever both switches have been pressed, otherwise make the LED labelled R glow.

(c) Explain how the program operates.

2 Here is a program segment.

```
       movi   s4, ED
loop:  dec    s4
       jnz    loop
```

(a) Explain how the segment provides a time delay.

(b) If each instruction is executed in 5 μs, calculate the time delay provided by the segment.

(c) Write a subroutine at label **pause** which provides a time delay of 1 ms.

3 This question is about the use of subroutines.

(a) Explain **two** advantages of using subroutines in programs.

(b) What are the **stack** and the **stack pointer**?

(c) Describe the changes in the stack and the stack pointer when a subroutine is used by the main program.

(d) Write a subroutine to provide a time delay of *n* ms, where *n* is a number stored in register s6. (Make use of the subroutine at **wait1ms** which provides a time delay of 1 ms.)

4 Write a subroutine which makes a microcontroller read in a word at the four lowest bits of the input port and copy it to the four highest bits of the output port, leaving the other four bits low. Explain how your subroutine operates.

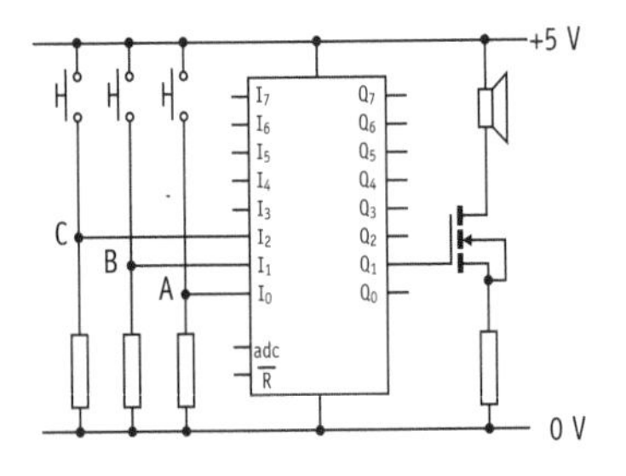

Fig 4.24 Circuit for question 5

5 This question is about the microcontroller circuit shown in Fig 4.24.

The system is programmed, to act as a musical keyboard with only three keys.

- Set the MOSFET gate high.
- Read in the word CBA into a register. This is the time delay in ms between changes of state of the gate.
- Use the subroutine at **wait1ms** to implement the delay.
- Use the EOR function to change the state of the gate and return to the second step.

Write a suitable program. Explain how the last step works.

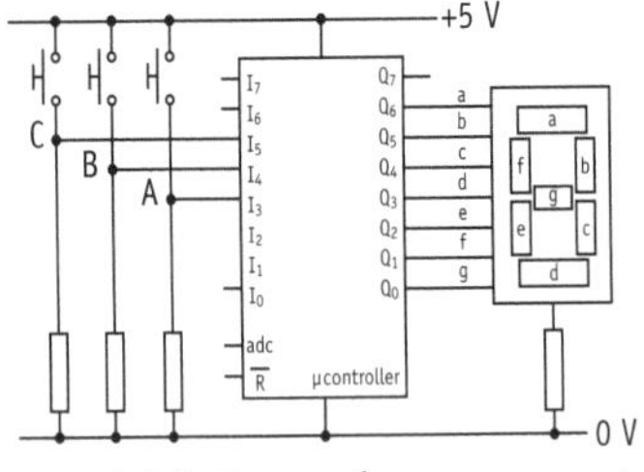

Fig 4.25 Circuit for question 6

6 This question is about the microcontroller circuit shown in Fig 4.25.

The program is to display the number CBA presented by the switches, coded in BCD.

(a) The program is to use a **look-up table** for the bytes to be placed at the output port. Draw up a suitable look-up table at the label **table**.

(b) Write the program, using the subroutine at **readtable**.

(c) Explain how the program operates.

Learning summary

By the end of this chapter you should be able to:

- understand the operation of a microcontroller as CPU, ports, clock and memory
- explain the use of the data, control and address buses in a microcontroller
- describe a machine cycle
- explain the role of the program counter in jumps
- understand how the stack and the stack pointer are used in subroutines
- use the AND operation to mask off unwanted data from the input port
- use the EOR operation to selectively change the state of bits at the output port
- use the subroutines at **wait1ms, readadc** and **readtable**
- write programs to provide time delays and use look-up tables.

CHAPTER

5 Video displays

5.1 Monochrome

Imagine looking at an image on a computer monitor screen. You are looking at a **video display**, an electronic system which mimics reality. It uses a continuous stream of information from the computer to paint what appears to be a moving picture on a flat surface. In fact, you are looking at a series of still pictures, each built up from a mosaic of tiny points, with each picture replaced by a different one quicker than your eyes can notice. The whole thing relies on the slowness and low resolution of your eyes and brain to trick you into seeing something which isn't really there.

Pixels

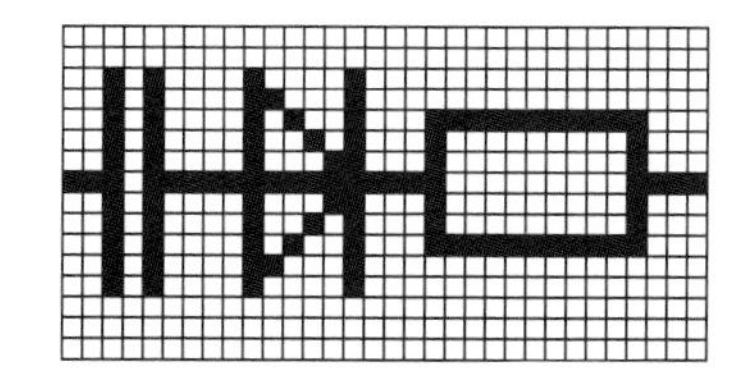

Fig 5.1 An image formed by a frame of 512 (16 × 32) pixels

The magic of video display starts with a **pixel**, a dot too small to be seen with the naked eye, which emits light. Arrange enough pixels into a **line** and stack enough lines one above the other, and you have a **frame** capable of displaying a useful image. For example, the frame of Fig 5.1 has 512 pixels arranged in 16 lines, with 32 pixels in each line. If you look at it from the other side of the room, the dots should merge into a single image.

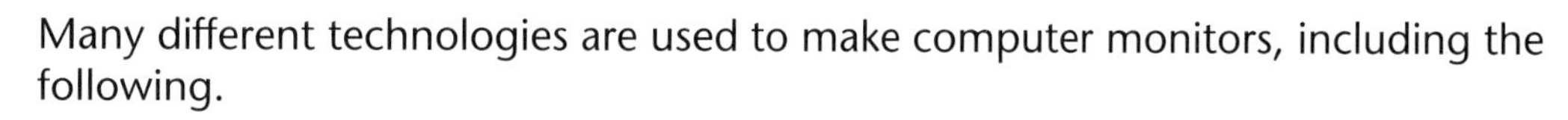

Many different technologies are used to make computer monitors, including the following.

- Cathode ray tubes (CRT) use a thin beam of electrons to make flashes of light on a phosphorescent screen.
- Liquid crystal displays (LCD) selectively allow light to pass through them, as used in flat screen monitors and digital projectors.

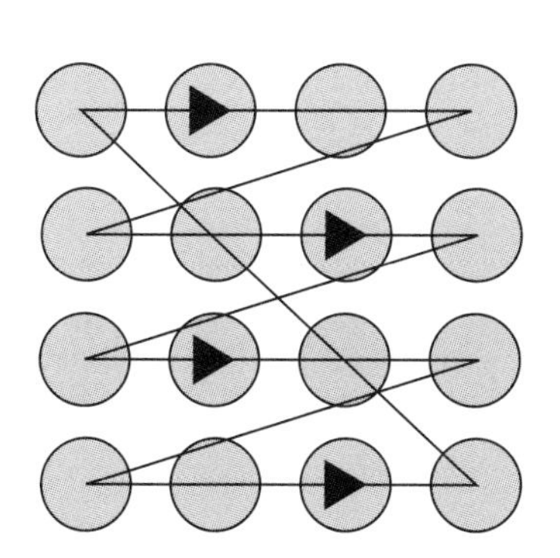

Fig 5.2 A raster scan for a 16-pixel screen

For many years, CRT screens were the only form of computer monitor. Since the electron beam can only arrive at one point on the screen at any one instant, CRT monitors employ a **raster scan** to display an image. This is illustrated in Fig 5.2.

Each pixel is turned on in turn, starting with the one at the top left-hand corner of the frame, moving from left to right along each line. In this way, the frame is built up line by line, starting from the top and ending at the bottom. Then the whole process is repeated. Provided that each frame is **refreshed** at least 25 times each second, the result is a flicker-free image. You don't notice that the image is built up from flashing dots.

Monitor signals

Three different signals need to be transferred from a computer to its monitor.

- The **video signal** controls the brightness of the pixels.
- The **line sync** signal instructs the monitor to start scanning a new line.
- The **frame sync** signal instructs the monitor to start scanning a new frame.

It makes sense for the same trio of signals to be used for both LCD and CRT monitors, even though LCD pixels remain on (once turned on) or off (once turned off), rather than being flashed on periodically. So in an LCD monitor the video signal is used to refresh each pixel, rather than just allow it to shine for a moment. Fig 5.3 shows how the three signals are related for a simple 16-pixel screen displaying an image.

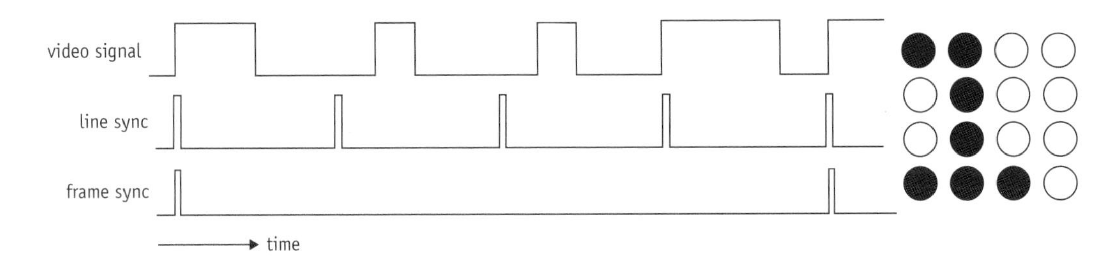

Fig 5.3 Timing diagram for the three computer monitor signals

Notice how the video signal changes with time. When it is high, the pixel selected at that instant in the raster scan emits light. When the video signal is low, the pixel is turned off. This means that the image is displayed in **monochrome**, as a single colour with just two levels of brightness (on or off).

Timing

The timescale of all three monitor signals in Fig 5.3 depends on the **frame refresh rate**. Suppose that it is 25 Hz, just fast enough to guarantee a flicker-free image. Calculate the time required for one raster scan first.

$T = ?$

$f = 25$ Hz

$$f = \frac{1}{T} \text{ so } T = \frac{1}{f} = \frac{1}{25} = 4\times10^{-2} \text{ s or } 40 \text{ ms}$$

If one frame takes 40 ms, then each of the four lines has 40 / 4 = 10 ms in the raster scan. Similarly, if one line takes 10 ms to scan, then each of the four pixels is controlled by the video signal for 10 / 4 = 2.5 ms. The value of this **access time** is important, especially if the technology used to make the screen means that pixels have a slow response time.

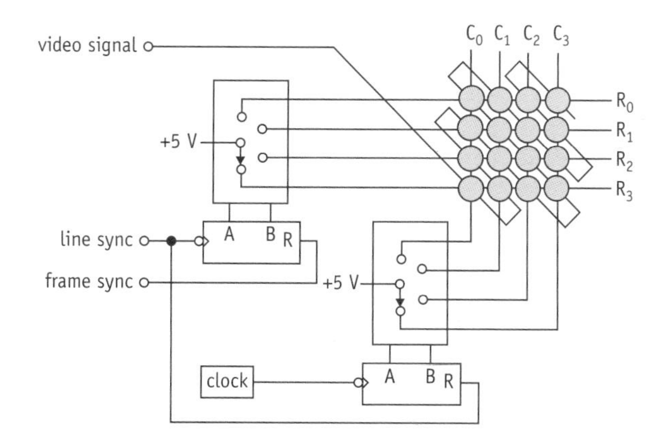

Fig 5.4 Each pixel is accessed by the video signal in turn

Accessing pixels

The system of Fig 5.4 shows a system which allows the video signal to access the pixels in the order required for a raster scan. Each of the 16 pixels is represented by a grey circle. Three signals reach each pixel:

- A row signal R which can be made high by a demultiplexer and binary counter triggered by pulses in the line sync signal.
- A column signal C which can be made high by another demultiplexer–counter pair which is triggered by pulses at 2.5 ms intervals from a 400 Hz clock.
- The video signal V which controls the state of whichever pixel has both R and C high.

You should be able to use the timing diagram of Fig 5.3 to work out the effect of the line sync and frame sync on the system. Notice, in particular, how each line sync pulse not only resets the scan to the first column of pixels, but also moves the scan down to the next row.

Pixel circuitry

Fig 5.5 shows two possible ways of building a pixel for the system of Fig 5.4.

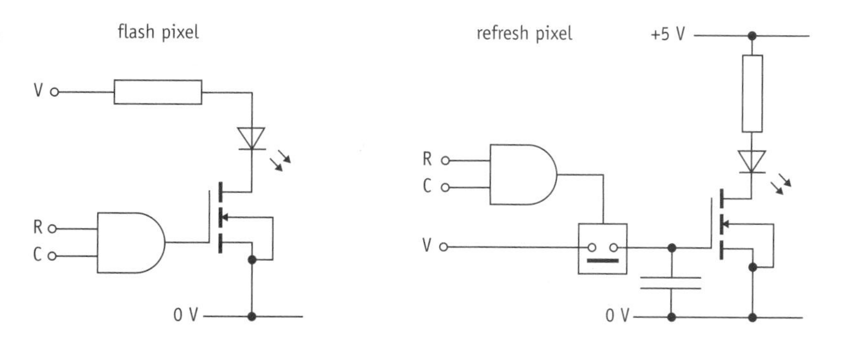

Fig 5.5 Two ways of using an LED as a pixel for Fig 5.4

The circuit on the left employs a MOSFET as a switch. Its gate can only go high when both R and C are high, allowing the video signal to flash the LED on for 2.5 ms. The circuit on the right uses a capacitor to keep the MOSFET gate high or low while the **analogue switch** is open. Only when both R and C are high does the switch close, allowing the video signal 2.5 ms to refresh the pixel by charging or discharging the capacitor.

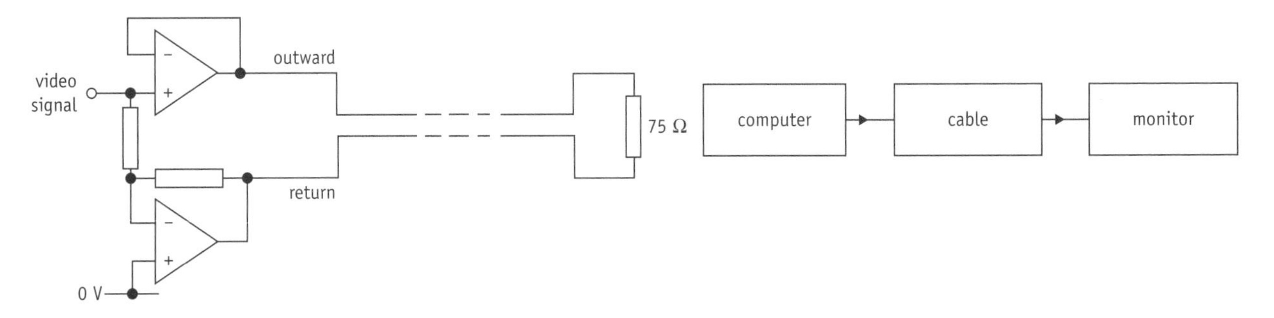

Fig 5.6 The video signal cable has two wires

Cable connection

Fig 5.6 shows that the cable which carries the video signal from a computer to its monitor contains two wires in parallel. The **outward** wire follows the video signal, and the **return** wire carries an inverted copy, with a 75 Ω resistor in the monitor to completely absorb the signal and prevent it from echoing back and forth along the cable. Of course, each wire has resistance, however small, and the two wires running in parallel look like a capacitor, suggesting the circuit of Fig 5.7.

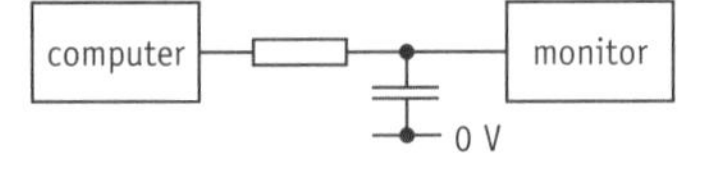

Fig 5.7 The cable behaves like a treble cut filter

The time constant *RC* of this equivalent circuit (a passive treble cut filter) will set a limit on the highest frequency that can be effectively transmitted by the cable. Only signals below the break frequency $f_0 = 1/2\pi RC$ will be passed from computer to monitor without any loss of amplitude. So the construction of the cable sets its **bandwidth**, the range of frequencies that it can transmit without any reduction. To see why this is an important consideration, consider the basic standard for computer monitors and data projectors.

- A frame refresh rate of up to 70 Hz.
- 480 lines of pixels in each frame.
- 640 pixels in each line.

This standard is called VGA, after the name of the circuit board (video graphics array) needed to implement it for early PC computers. You can estimate the necessary cable bandwidth in these four steps.

- Calculate the pixel access rate: $70 \times 480 \times 640 = 21\ 504\ 000$ pixels per second.
- Calculate the pixel access time.

  $T = ?$

  $f = 21\ 504\ 000$ Hz

  $$T = \frac{1}{f} = \frac{1}{21504000} = 4.7 \times 10^{-8} \text{ s or } 47 \text{ ns}$$

- Select the video signal which sets neighbouring pixels to opposite states. This will look like 1010101010, a square wave with a period of $2 \times 47 = 94$ ns.
- Calculate the frequency of the square wave.

$f = ?$

$T = 94$ ns

$$f = \frac{1}{T} = \frac{1}{94 \times 10^{-9}} = 1.1 \times 10^{7} \text{ Hz} \simeq 10 \text{ MHz}$$

As you will find out below, the lowest frequency sine wave required to make this square wave will therefore have a frequency of 10 MHz. The bandwidth of the cable needs to be greater than this if the video signal 10101010 is to pass from computer to monitor. A video signal which sets adjacent pixels on the screen to different states is called the **worst-case signal**. It has this name because it is the signal which requires the highest frequency to be sent down the cable.

Making waves

Fig 5.8 The worst-case video signal

amplitude	frequency
2.5 V	0 MHz
3.2 V	10 MHz
1.1 V	30 MHz
0.6 V	50 MHz
0.4 V	70 MHz

Any signal can be made by adding together **sine waves** of appropriate amplitude and frequency. It turns out that a **square wave**, such as the 10 MHz worst-case signal of Fig 5.8, requires an infinite number of different component sine waves to be added together, each with a different frequency and amplitude. The first five are listed in the table.

Fortunately, higher frequency components have smaller amplitudes than lower frequency ones, so you can usually manage with just the first two or three components. Voltage–time graphs for the first two components at 0 MHz and 10 MHz are shown in Fig 5.9. Adding those two signals together (Fig 5.10) gives a workable imitation of the square wave of Fig 5.8, certainly good enough to set the state of pixels which are too small to be seen individually.

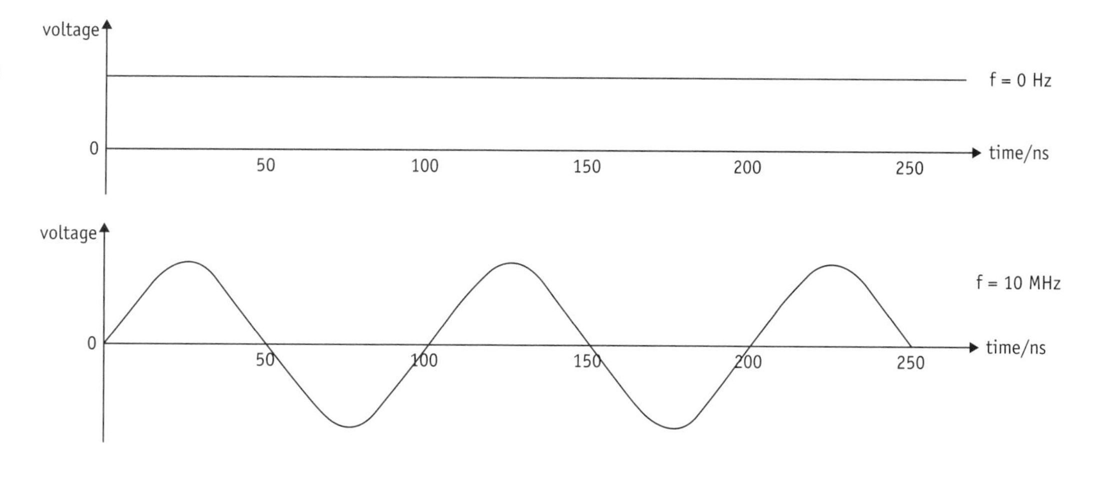

Fig 5.9 The largest components of the worst-case video signal

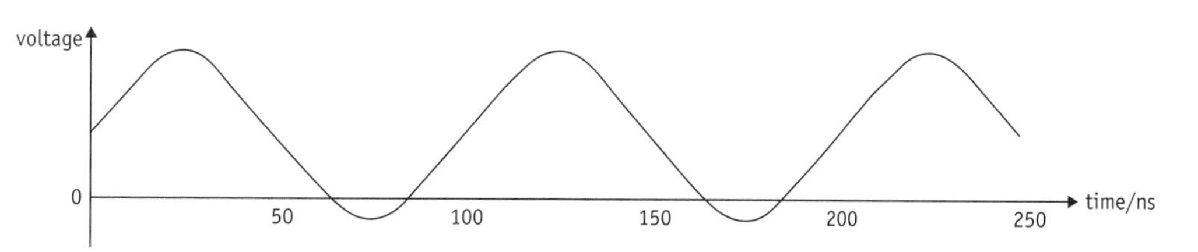

Fig 5.10 The sum of the two lowest frequency components

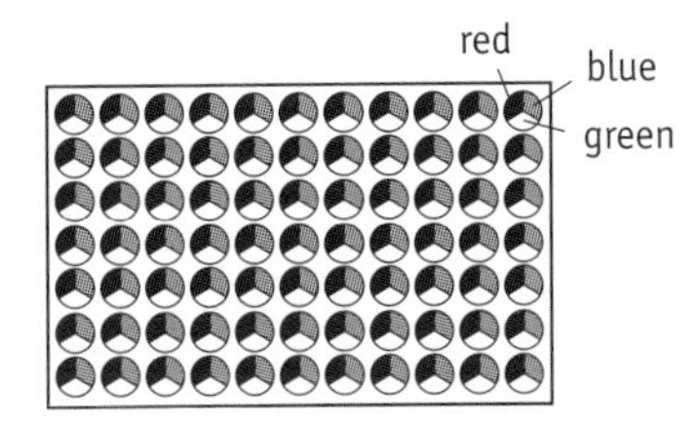

Fig 5.11 Each cluster has three different colours of pixel

5.2 Colour

Because of the way that our eyes are constructed, only three different coloured pixels (red, green and blue) are necessary to create all the colours that we can perceive. For example, red and green light together are perceived as yellow light. So screens that are designed to display colour images have their pixels grouped together, with one of each colour in each **cluster**, as shown in Fig 5.11. By turning each of the pixels in a cluster on or off, we can generate all of the colours shown in the table.

red pixel	green pixel	blue pixel	cluster colour
off	off	off	black
off	off	on	blue
off	on	off	green
off	on	on	cyan
on	off	off	red
on	off	on	magenta
on	on	off	yellow
on	on	on	white

Variable intensity

A range of only eight colours for a monitor screen is somewhat limiting. A much wider range can be achieved by allowing the video signal to vary the intensity of a pixel, not just turn it on and off. So each of the separate video signals for red, green and blue is **analogue**, not digital. This does not mean that the video signals can have any value within a range (usually 0 V to +1.4 V). The fact that each video signal is created by a computer means that it can only have a limited number of values. This is because computers manage data in digital form, as binary words. So if the intensity of a pixel is coded as a 2-bit word in the computer, the intensity can have only one of four intensity levels, as shown in Fig 5.12 and the table below.

binary word	video signal	pixel intensity
00	0.0 V	dark
01	0.4 V	dim
10	0.8 V	light
11	1.2 V	bright

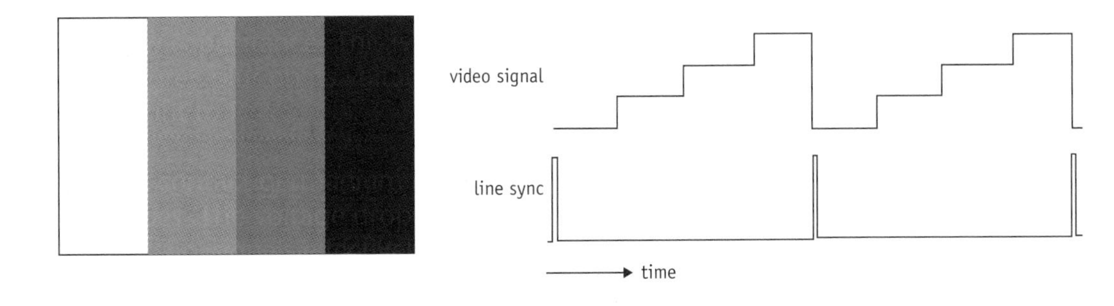

Fig 5.12 The effect of four different levels for the video signal

Digital-to-analogue conversion

Fig 5.13 shows a circuit which generates a different voltage at its output for each possible 2-bit word BA at its inputs. Since it converts information from a digital format to an analogue format, it is called a **digital-to-analogue converter (DAC)**.

Fig 5.13 A 2-bit DAC based on op-amps

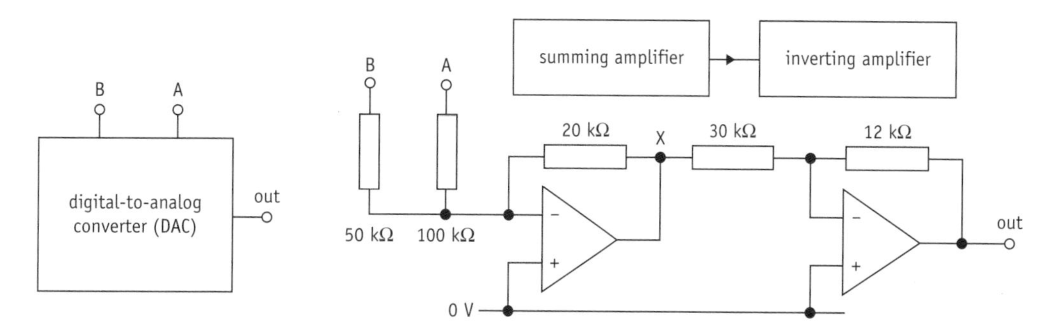

The block diagram shows that the system consists of a summing amplifier followed by an inverting amplifier. You will recall that the summing amplifier output is given this formula.

$$-\frac{V_{out}}{R_f} = \frac{V_1}{R_1} + \frac{V_2}{R_2}$$

Since V_1 is 5 V when B is 1 and 0 V when B is 0, $V_1 = 5B$. Similarly, $V_2 = 5A$. Put these into the equation with the resistor values in kilohms and you get this expression.

$$-\frac{V_X}{20} = \frac{5B}{50} + \frac{5A}{100}$$

Now multiply both sides by −20.

$$V_X = -(2B + A)$$

The table below gives the voltage at X for each of the four binary words BA.

BA	V_X	V_{out}
00	0 V	0 V
01	−1 V	0.4 V
10	−2 V	0.8 V
11	−3 V	1.2 V

The inverting amplifier has a gain of −0.4.

$G = ?$

$R_f = 12\ \text{k}\Omega$

$R_{in} = 30\ \text{k}\Omega$

$$G = -\frac{R_f}{R_{in}} = -\frac{12 \times 10^3}{30 \times 10^3} = -0.4$$

So $V_{out} = -0.4 \times V_X$, leading to the values shown in the table above.

More bits

b	*N*
1	1
2	4
3	8
4	16
5	32
6	64

The VGA computer monitor standard requires that each of the red, green and blue pixels have 64 different levels of intensity. The number of levels, *N*, is related to the length of binary word *b* by this formula.

$$N = 2^b$$

The table shows how the number of levels increases with increasing word length. In particular, that 64 different levels of a video signal requires a 6-bit word in the computer.

So the DAC must have six inputs, as shown in Fig 5.14, with the video signal at the output going from 0 V to +1.4 V as the input goes from 000000 to 111111.

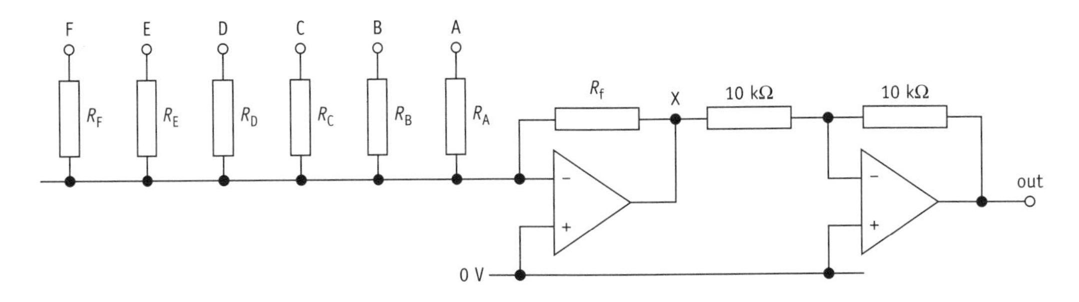

Fig 5.14 A six-input DAC

DAC design

So how do you decide on the resistor values for the DAC of Fig 5.14? There are five steps.

- Calculate the **resolution** of the DAC, the voltage difference between adjacent levels, from the number of levels (64) and the range of the video signal (1.4 V).

$$\text{resolution} = \frac{1.4}{64} = 2.2 \times 10^{-2}\ \text{V or 22 mV}$$

- Set the input resistor R_A for the **least significant bit (lsb)** A to some convenient large value, such as 256 kΩ.
- Make the value of R_B = 128 kΩ, exactly half that of R_A. This is because A represents 1 or 0, whereas B represents 2 or 0, so must deliver twice as much current towards the inverting input which sits at 0 V.
- Apply the same reasoning to set the values of R_C to 64 kΩ, R_D to 32 kΩ, R_E to 16 kΩ and R_F to 8 kΩ.
- Now set the input word to 000001, so that there is only current in R_A. The voltage at the output must then be equal to the resolution i.e. 22 mV. With current in only one resistor you can use the inverting amplifier formula to find the feedback resistor R_f.

$V_{out} = -1.4$ V

$V_{in} = +5$ V

$R_{in} = 256$ kΩ

$R_f = ?$

$$\frac{V_{out}}{V_{in}} = -\frac{R_f}{R_{in}} \qquad R_f = -R_{in}\frac{V_{out}}{V_{in}}$$

$$R_f = -256 \times 10^3 \times \frac{-1.4}{+5.0} = 7.2 \times 10^4\ \Omega \text{ or } 72\ \text{k}\Omega$$

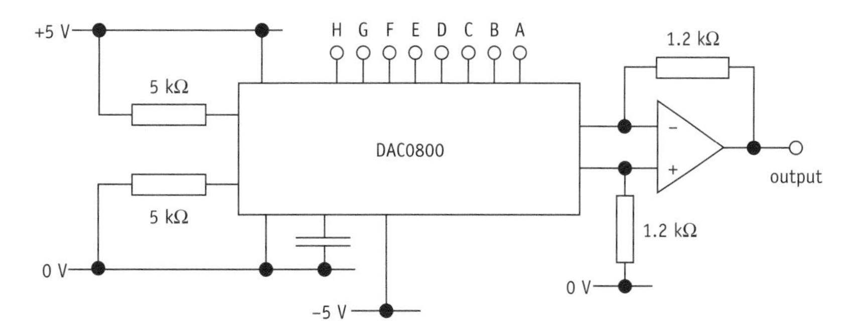

Fig 5.15 An integrated circuit DAC

Precision components

The values of the six input resistors of Fig 5.15 have to be just right for the circuit to work correctly. The tightest constraint is for R_F, the resistor for the **most significant bit** (or **msb**) of the word at the input. It needs to be 8 kΩ with a precision of at least 1 in 64. It is tricky to achieve this with discrete resistors, so DACs are usually constructed as integrated circuits. Fig 5.15 shows how the popular DAC0800 i.c. can be used with an op-amp to make a DAC which has a **range** of +1.2 V to −1.2 V for an 8-bit input. Since the DAC has $2^8 = 256$ different output levels, this means that it has a resolution of $2.4/256 \simeq 1\times10^{-2}$ V or 10 mV. The coding which relates the digital input to the analogue output is shown in the table.

digital input	analogue output / V
1111 1111	+1.20
1111 1110	+1.19
1111 1101	+1.18
...	...
0000 0010	−1.18
0000 0001	−1.19
0000 0000	−1.20

Digital versus analogue

As you will find out in later chapters, sending information from one place to another in analogue format leaves it wide open to corruption by noise and interference. Once the signal has arrived at its destination, there is nothing you can do to separate the signal from the noise and interference. Digital signals also pick up the same amount of noise and interference, but clever circuitry allows all the information to be successfully restored. The short distance from a monitor to a screen makes it unlikely that enough rubbish will get added to the red, green and blue video signals to make a visible difference on the screen. It is a different matter when video signals have to travel a long way, as they have to do in broadcast television (or **TV**). However, sending video information as a stream of bits has one major disadvantage. It requires a much larger bandwidth.

Bit rate

Consider the VGA monitor standard.

- 64 levels per pixel.
- 640 clusters of pixels in each line.
- 480 lines of clusters in each frame.
- A frame refresh rate of up to 70 Hz.

Suppose that you want to send all of this information as a single video signal in digital format. What **bit rate** would you need? Start off by considering the word length needed to set the brightness of a single pixel.

- $64 = 2^6$ so a 6-bit word is required for each pixel.
- A cluster of red, green and blue pixels will therefore need an 18-bit word.
- The bits required for each line are $18 \times 640 = 11\,520$.
- The bits required for each frame are $11\,520 \times 480 = 5\,529\,600$.
- So the total bit rate is $5\,529\,600 \times 70 = 387\,072\,000$ bits per second.

In practice, the bit rate needs to be a bit higher to include synchronisation signals, taking it to a round figure of 400 000 000 bits per second, or 400 Mbit s^{-1}.

Bandwidth

Bit rate is not the same as bandwidth. To determine the bandwidth from the bit rate, you need to consider the **worst case signal**, one whose basic components have the highest frequency. For a digital signal, this is 101010101, with bits alternating between 1 and 0. So the worst case digital signal at 400 Mbit s^{-1} looks like a square wave oscillating at 200 MHz. (Each 1−0 pair makes up one cycle of the square wave.) Of course, a square wave at 200 MHz requires the combination of many different sine waves. This is shown in the **frequency spectrum** of Fig 5.16.

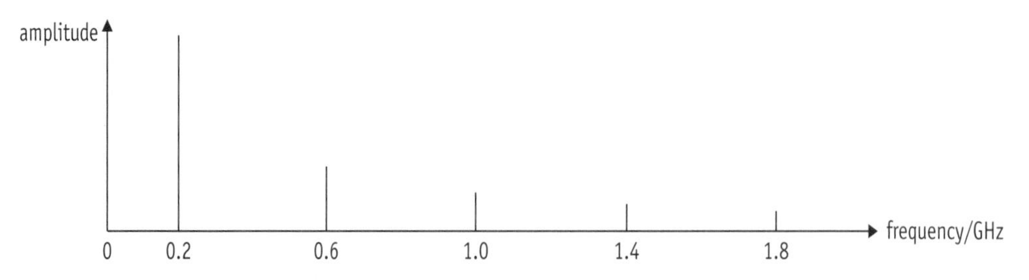

Fig 5.16 Frequency spectrum of a 200 MHz square wave

A frequency spectrum is an amplitude–frequency graph for a signal. It shows the relative amplitudes of the various **component** sine waves needed to make up the signal. In this case, a square wave at 200 MHz (or 0.2 GHz) needs components at 0.2 GHz, 0.6 GHz, 1.0 GHz . . . with decreasing amplitude as the frequency increases. However, you only need to transmit the lowest frequency component because a logic gate easily restores all of the other components upon arrival, as shown in Fig 5.17. So the transmission bandwidth required is only 200 MHz.

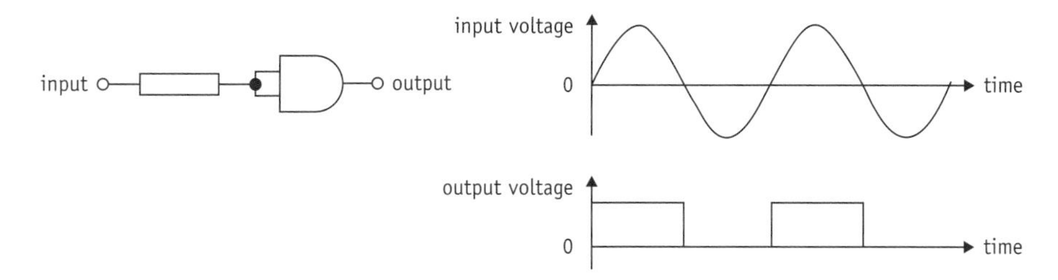

Fig 5.17 Using a logic gate to restore digital signal

The following rule can be applied to all situations where restoration of a square wave from a sine wave is possible.

bandwidth (in Hz) = 0.5 × bit rate (in s^{-1})

Compression

200 MHz is a very large bandwidth. To put it in perspective, consider the storage capacity of a standard CD-ROM. This is almost 6 Gbits. If a 200 MHz bandwidth delivers 400 Mbits in each second onto a CD-ROM, it will be full after only 6/0.4 = 15 s ! As you know, a CD-ROM can easily hold an hour's worth of full colour video information (and allied sound track). This is because the digital video stream is **compressed** before it is stored on the disk.

There are many ways in which a digital signal can be compressed. The most widely used one for video is called MPEG, named after the group of people who devised it (the Moving Pictures Expert Group). It uses more than one technique to reduce the number of bits per frame. Here are two of them.

- The red, green and blue signals for each cluster are recoded as a **luminance** signal and a pair of **chrominance** signals. The luminance signal is essentially the sum of the red, green and blue signals, giving the brightness of the cluster. The chrominance signals are red plus green and green plus blue, and between them they hold all of the information about the colour of the cluster. Now, the human eye is less sensitive to colour changes than it is to brightness changes, so the chrominance signals can manage with fewer levels than the luminance one. So the 18 bits normally required for the pixels of the cluster could be recoded as a 6-bit luminance signal and a pair of 2-bit chrominance signals, giving only 10 bits – a reduction of almost 50%.
- There is often only a small difference between adjacent frames, so that change can be encoded with far fewer bits than are required for a completely new frame. So if you send an entire frame, you don't have to follow it with others – just send the changes required to update each frame to make it into the next one. In fact, MPEG requires that every 15th frame be sent in full, accompanied by the changes for the intervening 14.

Both of these compression techniques result in a loss of information from the original three video signals (red, green and blue). This means that the quality of the final image is not as good as the original, so a balance needs to be struck between reducing the bandwidth to a manageable level and producing a picture on the screen which is acceptable.

Questions

5.1 Monochrome

1 An XGA monitor uses a raster scan to display **frames** of 786 432 pixels with a **refresh rate** of 60 Hz.

(a) Explain the meaning of each of the terms in bold.

(b) Each frame contains 768 lines. Calculate the number of pixels in each line.

(c) Explain why the bandwidth of the cable leading to the monitor needs to be at least 24 MHz.

2 Three different signals pass from a computer to its monitor. One of them is the video signal.

(a) State the function of the video signal.

(b) State the names and functions of the other two signals.

(c) Explain how the three signals allow the image to be built up on the screen.

(d) Explain why the refresh rate of a monitor needs to be at least 25 Hz.

3 Fig 5.18 gives timing diagrams for the line sync and frame sync signals of a CGA monitor.

Fig 5.18 Question 3

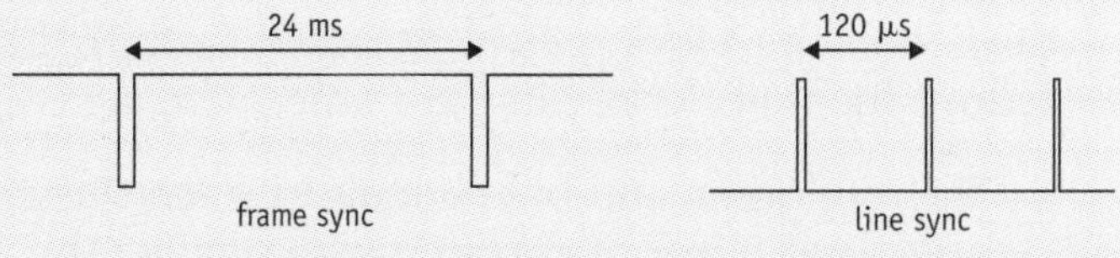

(a) Explain how the timing diagrams show that there are 200 lines of pixels in each frame.

(b) Calculate the refresh rate.

(c) The bandwidth required to pass the video signal from the computer to the monitor is 1.3 MHz. Calculate the number of pixels in each line.

5.2 Colour

1 Five different signals pass from a computer to its monitor.

(a) Describe the function of the two synchronisation signals.

(b) Describe the function of the three video signals.

(c) Explain why the synchronisation signals can be digital, but the video signals are analogue.

2 An EGA monitor has four different levels of intensity for each of the three pixels in a cluster.

There are 640 clusters of pixels in each line, with 350 lines in each frame.

The refresh rate is 60 Hz.

(a) Why are three pixels required for each cluster?

(b) Show that a 6-bit word can hold the information for controlling a single cluster.

(c) Calculate the total bit rate required to paint a moving picture on the screen.

(d) Digital video signals are usually compressed. Explain what is meant by **compression**.

(e) State the advantages and disadvantages of compressing a digital video signal.

3 The table shows how the voltage V_{out} at the output of a particular DAC is related to the word DCBA at its inputs.

DCBA	V_{out} /V
1111	3.00
1110	2.80
...	...
0001	0.20
0000	0.00

(a) State the range and resolution of the DAC.

(b) Show how a summing amplifier and an inverting amplifier could be used to make the DAC. Show all component values and justify them with calculations.

4 A colour monitor has 768 lines, with 1024 clusters of pixels in each line. Each of the three video signals has eight different voltage levels.The refresh rate is 50 Hz.

(a) Calculate the bandwidth required of the cable which carries each of the analogue video signals.

An alternative method sends the video signals in digital format.

(b) What is the advantage of sending the video signal in digital format?

(c) Show that the bandwidth required for each digital signal will be about 60 MHz.

(d) Suggest why video signals are sent from computer to monitor in analogue format.

CHAPTER

6 Modulating carriers

6.1 Amplitude modulation

Fig 6.1 An early electric telegraph transmitter

In 1844, the electric telegraph was used for the first time to send information beyond the horizon. The transmitter (Fig 6.1) and the receiver (Fig 6.2) were joined by a pair of wires. The transmitter was just a switch which connected a battery to the receiver through the wires. The receiver contained an electromagnet, and each time the switch was closed, the electromagnet pulled a pen onto a moving strip of paper. So by rapidly opening and closing the switch, a series of dots and dashes could be made to appear on the paper. Provided that the battery had a high enough voltage, messages in an agreed code could be sent over hundreds of miles along pairs of wires.

Fig 6.2 An early electric telegraph receiver

The circuit of Fig 6.3 shows the circuit diagram of an old electric telegraph. The table shows how the current in the wires is determined by the state of the switch. The switch changes (or **modulates**) the current to convey information. The system transmits information along the wires by altering the current in them.

Switch	Current
open	off
closed	on

The code used to convey information along the wires had only two values:

- A short burst of current in the wires (or **dot**).
- A long burst of current in the wires (or **dash**).

Dots and dashes were then strung together to represent numbers or letters. For example, a dot followed by a dash represented A, a dash followed by three dots represented B, and so on. This use of **Morse code** to convey messages across continents is not unlike the way that computers use the internet today to exchange emails. Both employ a **digital code** which has only two symbols (dot/dash or 1/0) which get strung together in different combinations to represent things. The only aspect which has really changed is the transmission speed. A typical Morse code operator could sustain a transmission rate of only one letter per second!

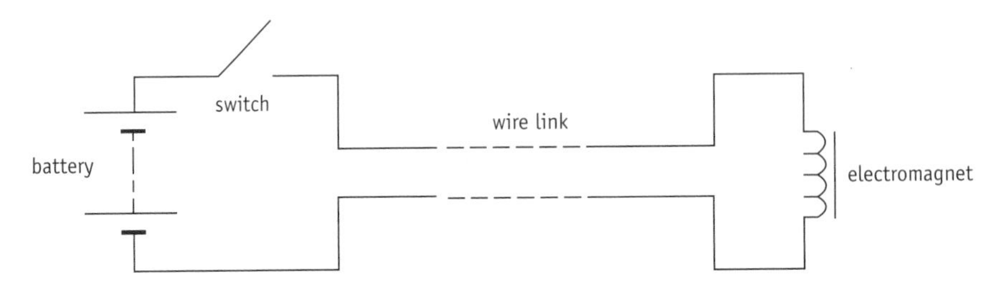

Fig 6.3 Circuit diagram for an electric telegraph system

Fig 6.4 Early telephone

Digital disadvantages

The Morse code telegraph suffered from one major disadvantage: it was slow. Consider the transmission of someone's spoken message through the telegraph. The following steps were required.

- Write down the message using letters of the alphabet.
- Use Morse code to transmit the letters one after the other.
- Decode the long and short pulses of current into letters.
- Read out the received message.

It should be obvious that not only did this process take a lot of time, but also that the sound of the final received message was nothing like that of the original. The coding process meant that only part of the original signal got through the system. This is a failing of all digital transmission schemes which have to transmit **analogue** signals (such as sound) which need more than two levels.

Analogue advantages

The telephone (Fig 6.4) was the first communication system to use analogue modulation of current in wires to transmit information. Fig 6.5 is the circuit diagram for a simple one-way telephone system.

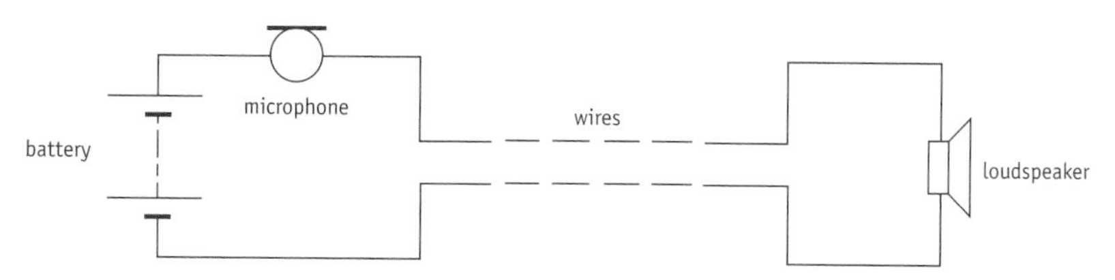

Fig 6.5 Circuit diagram for an electric telephone system

As with the electric telegraph (Fig 6.3), there is a component to modulate the current in the connecting wires. The microphone is essentially a resistor whose value depends on the air pressure around it. So as a sound wave passes over it, the resistance goes up and down, modulating the current in the wires. Fig 6.6 shows a current–time graph that you would get if the sound at the microphone is a pure tone, containing a single frequency (440 Hz).

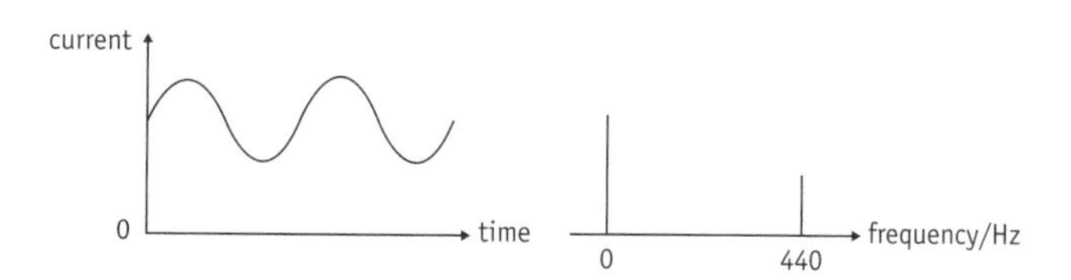

Fig 6.6 Current–time graph and frequency spectrum for a telephone signal

The frequency spectrum of Fig 6.6 shows that the current in the wires has a d.c. component (at 0 Hz) and an a.c. component (at 440 Hz). The loudspeaker at the other end of the wires uses the a.c. component of the current in it to make sound. In this way, a respectable copy of the sound at the transmitter can be recreated at the receiver in real time. There is no delay for coding and decoding the information, and all of the information is transmitted. However, as with all analogue transmission systems, the telephone's range is limited by noise. As the distance between transmitter and receiver is increased, the sound from the loudspeaker is increasingly dominated by the hiss from signals picked up by the wires.

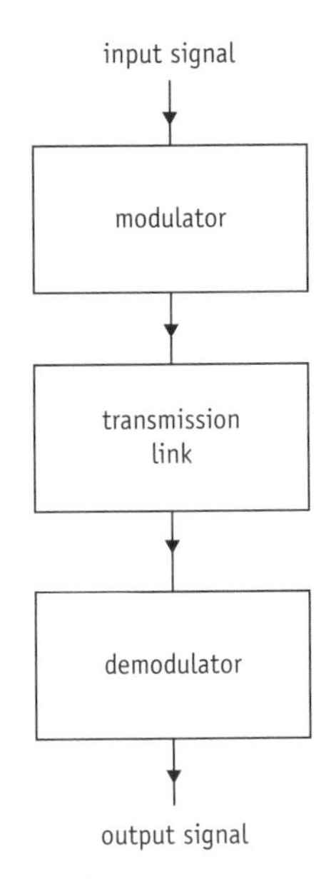

Fig 6.7 Block diagram for a signal transmission system

Modulate, demodulate

Any communication system which transfers information from one place to another along wires can be described by a block diagram like that of Fig 6.7.

Each block has a different function.

- The **modulator** codes the signal by varying the current at its output.
- The **transmission link** is the pair of wires which carry the current.
- The **demodulator** extracts the signal from the varying current.

As we saw on the previous page, there is more than one way of modulating a signal. The rest of this chapter will introduce you to three different ways of modulating an analogue signal. None of them is perfect.

Variable gain

Fig 6.8 shows an amplitude modulator.

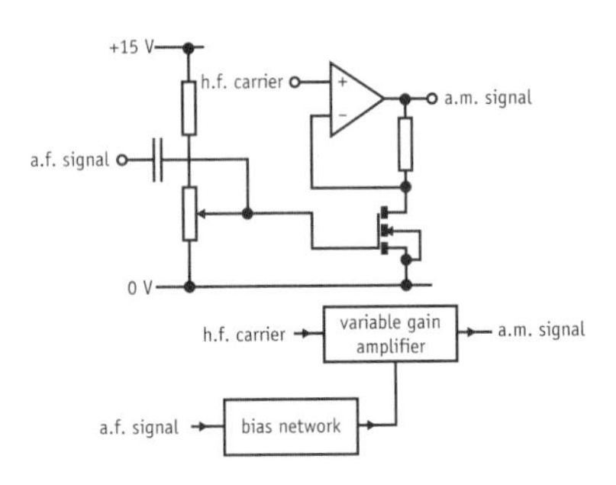

Fig 6.8 Amplitude modulator based on a variable gain amplifier

There are two distinct sub-systems:

- The **bias network** of resistor, potentiometer and capacitor adds a variable d.c. component to the **audio frequency** (or **a.f.**) **signal**.
- The **variable gain amplifier** uses the signal from the bias network to alter the amplitude of the **high frequency** (or **h.f.**) **carrier** at its input.

The function of the whole system is illustrated in the three graphs of Fig 6.9.

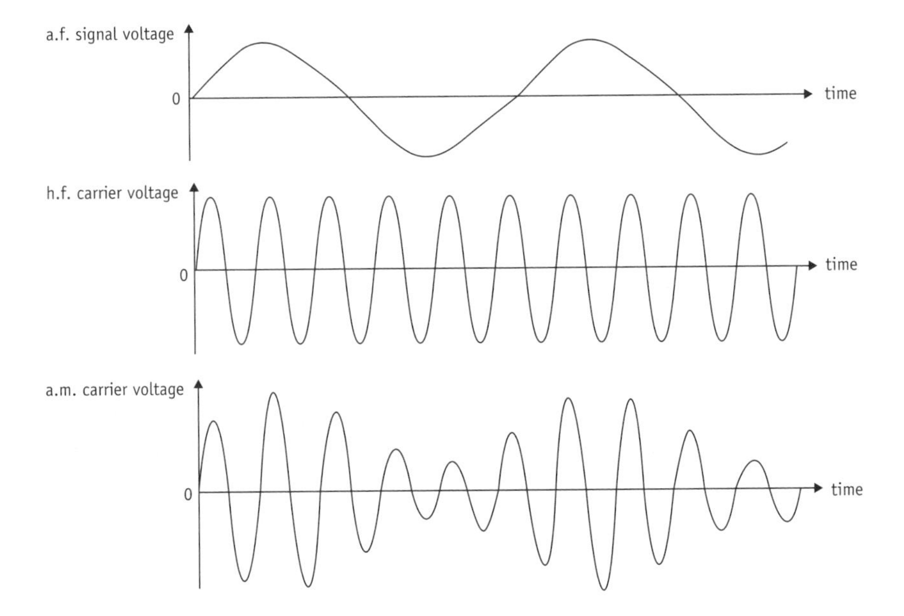

Fig 6.9 Voltage–time graphs for signals at the inputs and output of an amplitude modulator

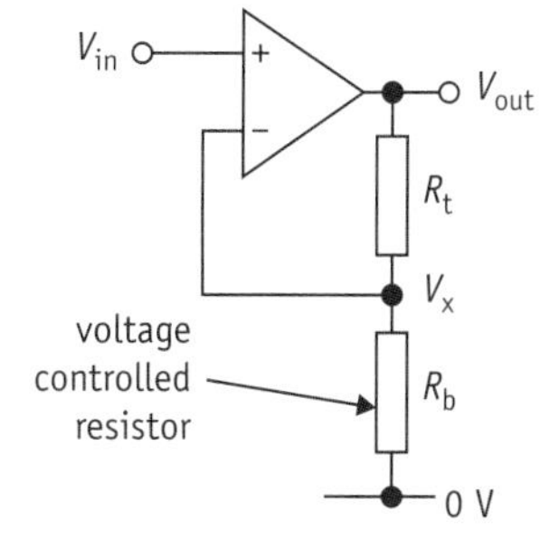

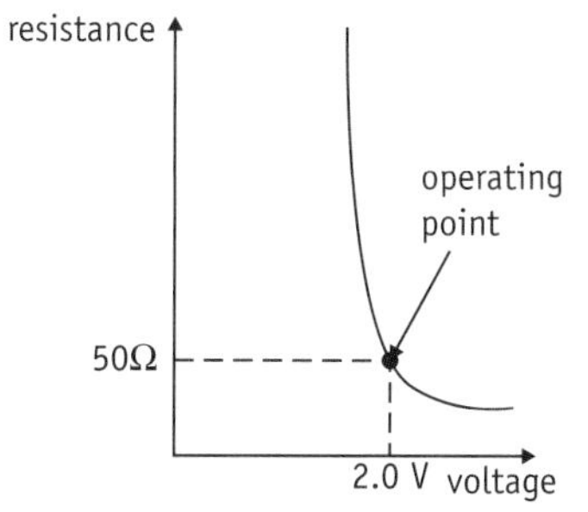

Fig 6.10 The MOSFET is used as a voltage-controlled resistor

For simplicity, the a.f. signal has been drawn as a sine wave – a pure tone with a single frequency. In reality, an a.f. signal will contain a range of frequencies between, say, 20 Hz and 20 kHz, the range of human hearing. The h.f. carrier is at a much higher frequency. The **amplitude modulated** (or **a.m.**) carrier is essentially an amplified copy of the h.f. carrier, where the gain has been controlled by the moment-by-moment voltage of the a.f. signal. So where the a.f. signal has its maximum positive value, the amplitude of the a.m. carrier is at its largest, and where the a.f. signal has its maximum negative value, the amplitude of the a.m. carrier is at its lowest.

MOSFET behaviour

So how does the variable gain amplifier operate? As you can see from Fig 6.10, it is essentially an op-amp connected as a non-inverting amplifier, with one fixed resistor (R_t) and one voltage-controlled resistor (R_b). The latter is actually a MOSFET, whose drain–source resistance decreases with increasing gate–source voltage, as shown in the graph. From your work at AS, you will recall the formula for the voltage G of a non-inverting amplifier:

$$G = 1 + \frac{R_t}{R_b}$$

The MOSFET needs to be biased so that any change of gate voltage results in a change of resistance. The bias required to obtain a satisfactory operating point will vary according to the type of MOSFET used. The graph of Fig 6.10 assumes that the drain–source resistance is 50 Ω when the gate is held 2.0 V above the source. This means that using 470 Ω for R_t gives the amplifier a gain of about +10 at the operating point.

$G = ?$

$R_t = 470\ \Omega$

$R_b = 50\ \Omega$

$$G = 1 + \frac{R_t}{R_b} = 1 + \frac{470}{50} = 10$$

The amplitude of the h.f. carrier at the non-inverting input needs to be quite small for the MOSFET to behave like a resistor. Again, this depends on the type of MOSFET being used, but typically it needs to be less than 100 mV. Let's choose 50 mV, so that the a.m. carrier has an amplitude of 500 mV at the operating point. Any a.f. signal applied to the modulator is added to the 2.0 V bias signal, changing the gain as the MOSFET resistance increases and decreases. The amplitude of the a.m. carrier changes accordingly, going up and down as the voltage of the a.f. signal goes up and down.

Limitations

The use of this type of variable gain amplifier as an amplitude modulator has the following limitations.

- There is an upper limit of about 100 kHz on the carrier frequency. This is because most op-amps aren't much good at amplifying signals above 1 MHz.
- The amplitude of the a.f. signal needs to be quite small to avoid distortion of the information placed on the carrier by the modulation process. This is because the amplifier gain may not change linearly with the voltage of the a.f. signal – it all depends on the transfer characteristic of the MOSFET.

Both of these are overcome by a different circuit, shown on the next page.

Chopping

The a.m. modulator circuit of Fig 6.11 uses a different type of variable gain amplifier. The MOSFET and 1 kΩ resistor make up a switch which has a gain of either 1 or 0, depending on the voltage at the gate. When the MOSFET gate is high, the drain-source resistance is almost zero, forcing D down to 0 V. However, when the gate is low, the drain-source resistance is infinite allowing D to have the same voltage as the op-amp output A. As you can see from Fig 6.11, this switch is controlled by the output of a relaxation oscillator operating at about 500 kHz.

$f = ?$

$R = 42\ \text{k}\Omega$

$C = 100\ \text{pF}$

$$f = \frac{2}{RC} = \frac{2}{42 \times 10^3 \times 100 \times 10^{-12}} = 4.8 \times 10^5\ \text{Hz}$$

The square wave at the output of this oscillator is the h.f. carrier.

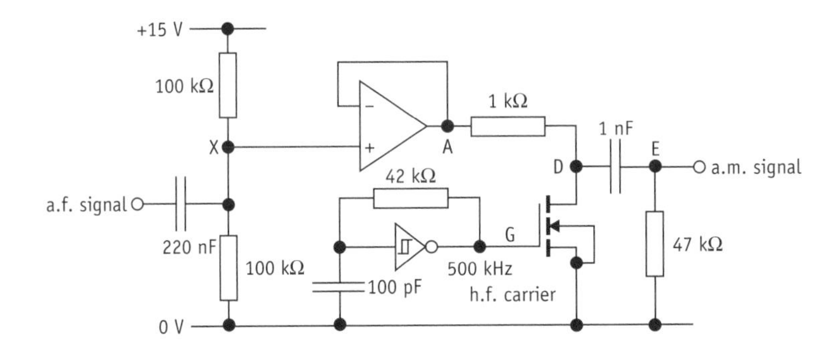

Fig 6.11 An a.m. modulator which uses a MOSFET as a switch

The op-amp is configured as a voltage follower, copying the signal at X to A. This allows the two 100 kΩ resistors of the bias network to present a usefully large input impedance of 50 kΩ to the source of the a.f. signal. The bias network and the coupling capacitor make up a bass-cut filter with a break frequency at the bottom end of the human range of hearing.

$f_0 = ?$

$R = 50\ \text{k}\Omega$

$C = 220\ \text{nF}$

$$f_0 = \frac{1}{2\pi RC} = \frac{1}{2\pi \times 50 \times 10^3 \times 220 \times 10^{-9}} = 14\ \text{Hz}$$

As you can see from the block diagram of Fig 6.12, there is another passive bass-cut filter at the output of the switch. This time, the break frequency is set to the upper end of the frequency range of the a.f. signal.

$f_0 = ?$

$R = 47\ \text{k}\Omega$

$C = 1\ \text{nF}$

$$f_0 = \frac{1}{2\pi RC} = \frac{1}{2\pi \times 47 \times 10^3 \times 1 \times 10^{-9}}$$
$$= 3.4 \times 10^3\ \text{Hz or } 3.4\ \text{kHz}$$

Although this is not the top end of the full range of human hearing, it is high enough for speech to be transmitted acceptably.

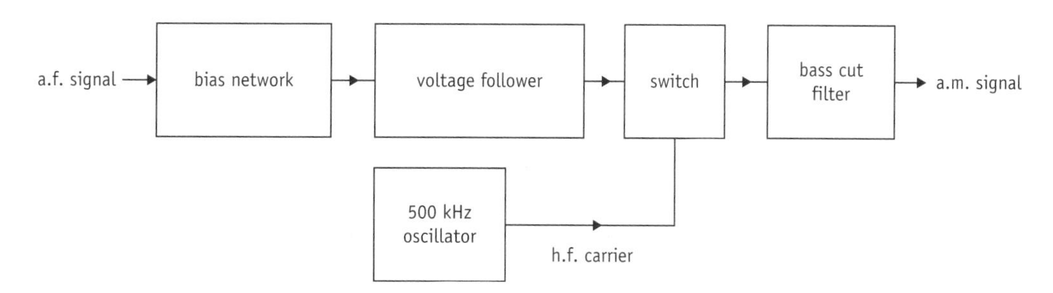

Fig 6.12 Block diagram for the modulator of Fig 6.11

Bias, chop and filter

The voltage–time graphs of Fig 6.13 show three stages in the process of copying the information in the a.f. signal onto the a.m. carrier.

- The bias network adds a d.c. component of 7.5 V, giving the signal at A.
- The h.f. carrier at G continually and rapidly turns the switch on and off, resulting in a chopped signal at D.
- The bass-cut filter removes the remaining a.f. components at D to give the a.m. carrier at E.

Since the MOSFET is being used as a switch, you don't have to worry about distortion due to its non-linear transfer characteristic. Nor does the op-amp impose an upper limit on the carrier frequency as it only has to handle the a.f. signal. For both of these reasons, the switched a.m. modulator of Fig 6.11 is much better than the one with continuously variable gain shown in Fig 6.8.

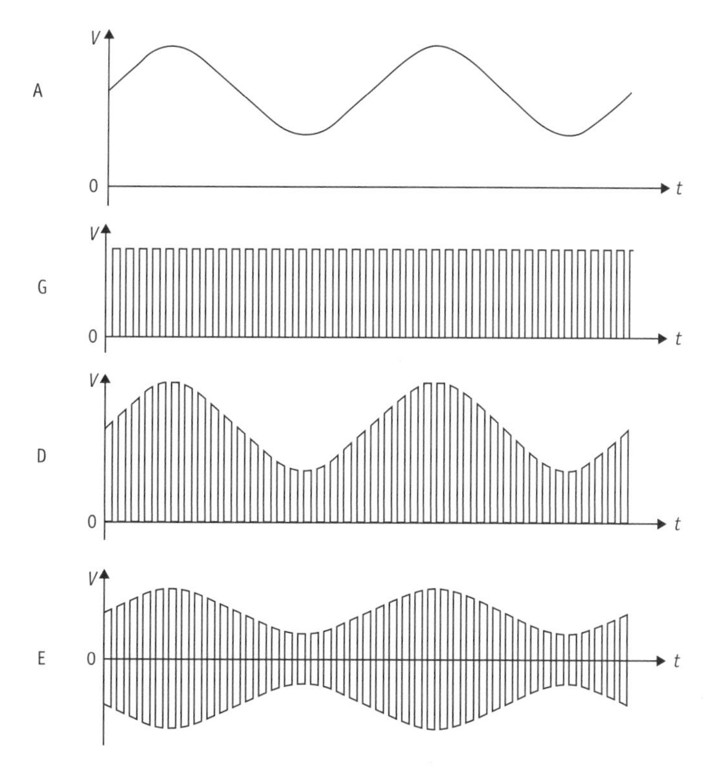

Fig 6.13 Signals at four places in the circuit of Fig 6.11

Frequency spectrum

The effect of modulating a carrier is easy to follow if you use frequency spectra instead of voltage–time graphs. For example, what is the result of modulating a 2 kHz a.f. signal onto a 10 kHz h.f. carrier? All is revealed in Fig 6.14.

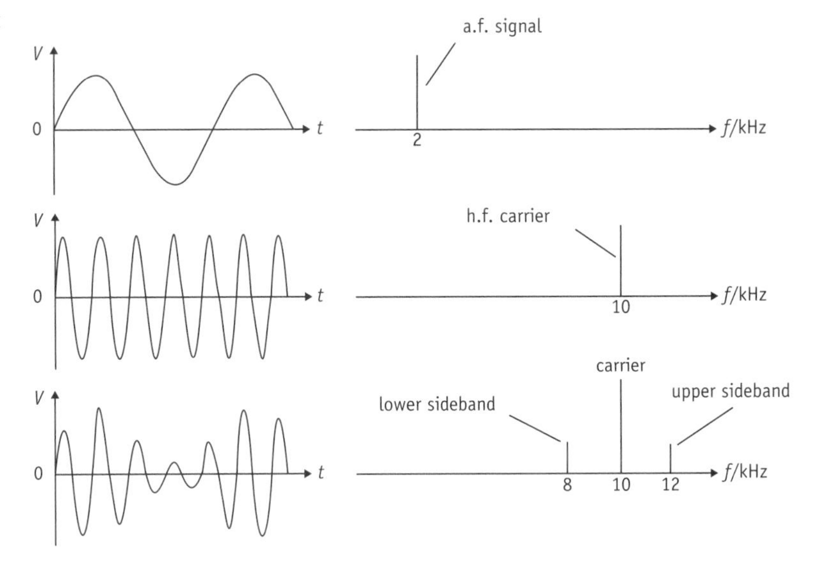

Fig 6.14 Frequency spectra of signals in a modulator

Three voltage–time graphs are shown, one each for the a.f. signal, the h.f. carrier and the resulting a.m. carrier. Next to each graph is its frequency spectrum, showing the various frequency components of the signal. In particular, notice that the frequency spectrum of the a.m. carrier contains three components.

- A large component at 10 kHz, the **carrier**.
- A smaller component at 8 kHz, the **lower sideband**.
- Another smaller component at 12 kHz, the **upper sideband**.

Only the sidebands contain information about the a.f. signal.

- The amplitude of each sideband is half that of the a.f. signal.
- The upper sideband frequency is that of the h.f. carrier plus the a.f. signal.
- The lower sideband frequency is that of the h.f. carrier minus the a.f. signal.

Adding components

So why does modulating a carrier create a pair of sidebands? The voltage–time graphs on the next page show, in two stages, how the addition of components at the relevant frequencies results in a signal with the correct shape. Fig 6.15 shows how the two sidebands at 8 kHz and 12 kHz add together over a time slice of 1 ms. Notice how they reinforce in some places and cancel out completely in others. The resulting signal has an amplitude which increases and decreases with the correct period but doesn't quite have the right shape. Only when a substantial amount of h.f. carrier at 10 kHz is added, as shown in Fig 6.16, is the correct waveform achieved.

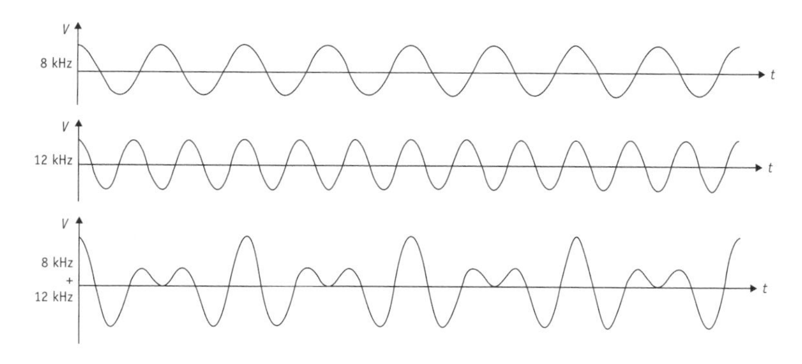

Fig 6.15 Adding sine waves at 8 kHz and 12 kHz

Bandwidth

You should now appreciate that the process of amplitude modulation simply creates two copies of the a.f. signal, each at a much higher frequency than the original, but with half the amplitude. The frequencies of these sidebands are given by a pair of simple rules.

$$f_{upper} = f_{carrier} + f_{signal}$$

$$f_{lower} = f_{carrier} - f_{signal}$$

This means that the bandwidth Δf_{am} needed to transmit an a.m. carrier is determined by the highest frequency f_s of the a.f. signal.

$$\Delta f_{am} = 2f_s$$

For example, a range of 300 Hz to 3400 Hz is commonly used in telephone systems for voice messages. What is the bandwidth required for an a.m. carrier which has been modulated by a voice message?

$\Delta f_{am} = ?$ $\quad$ $\Delta f_{am} = 2f_s = 2 \times 3400 = 6800$ Hz or 7 kHz

$f_s = 3400$ Hz

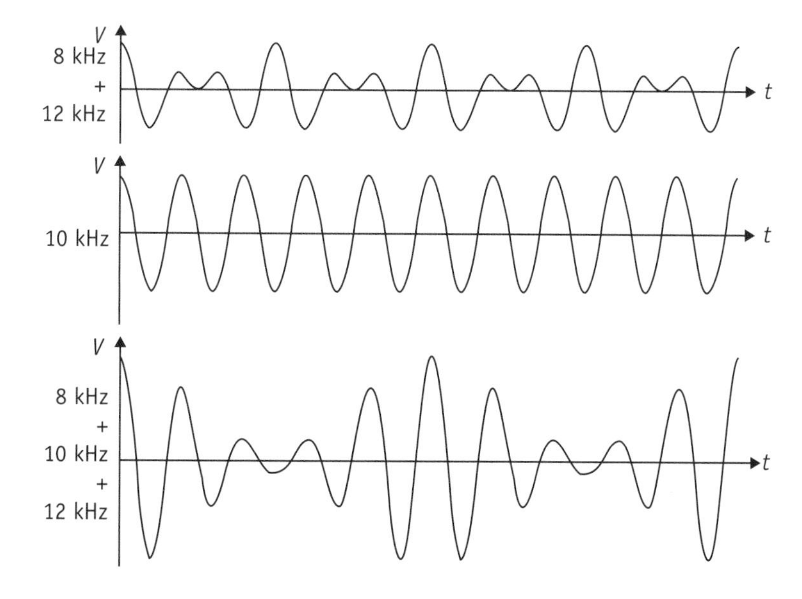

Fig 6.16 Adding all three components of the a.m. carrier

Demodulation

Demodulating an amplitude modulated carrier is much easier than modulating it in the first place. At its simplest, it only requires three components.

- A diode and resistor to rectify the a.m. carrier and recreate the a.f. signal.
- A capacitor in parallel with the resistor to filter out the carrier and sidebands.

The first stage is shown in the circuit of Fig 6.17.

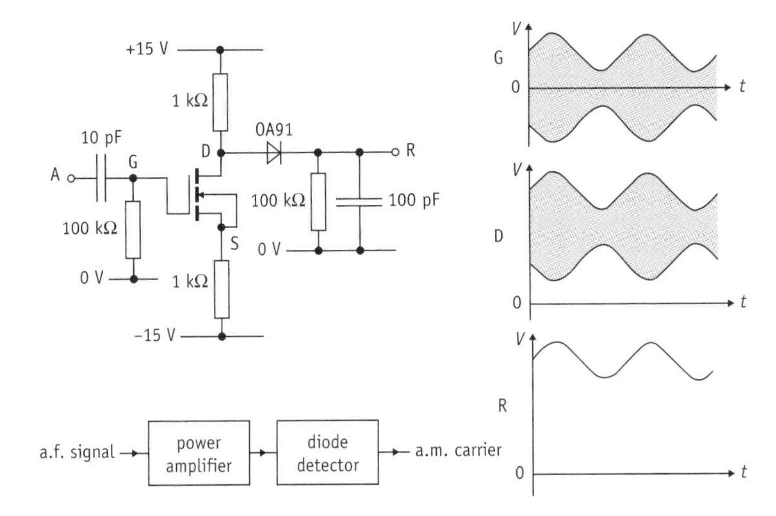

Fig 6.17 Diode detector for a.m. demodulation

As you can see from the block diagram, the circuit has two separate parts.

- The MOSFET power amplifier copies the a.m. carrier at A to D, adding a d.c. component of about 2 V, slightly more than the threshold voltage of the MOSFET. This allows the source of the a.m. carrier to see a load of 100 kΩ, yet presents an output impedance of only 1 kΩ to the next stage.

- The diode detector makes R the same as the peak value of D.

So an a.m. carrier applied at the input A results in the a.f. signal appearing at R. The information stored in the two sidebands of the carrier is used to make a copy of the original modulating signal.

Demodulation diodes

You can't use just any old diode in a useful a.m. demodulator. It has to have two important properties.

- A low voltage drop in forward bias. This is because a.m. carriers often have small amplitudes, so you can't afford to lose the 0.7 V of a typical silicon diode.

- A low capacitance. This is so that the high frequency a.m. carrier at D (shown as a grey blur on its voltage–time graph) can't bypass the diode and go straight to R.

A point-contact germanium diode (such as the OA91) has both of these properties. The voltage drop is only 0.2 V for a forward current of 0.1 mA, and it rectifies a.c. signals effectively up to about 100 MHz.

Diode detector

Fig 6.18 shows the flow of charge in the diode detector as the voltage at the input rises and falls rapidly.

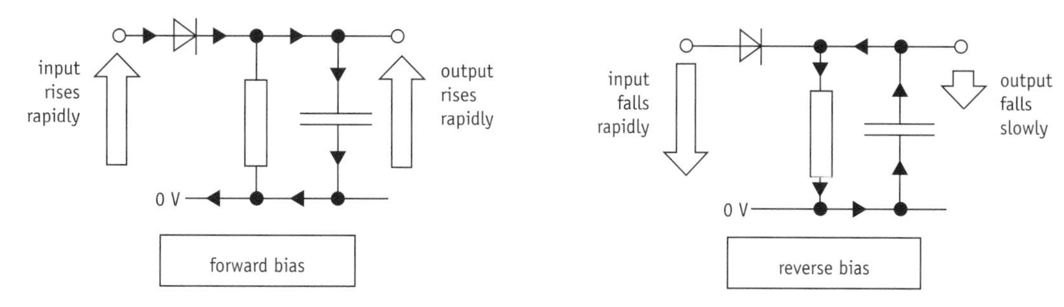

Fig 6.18 The diode is in forward or reverse bias

When the voltage at the input rises, the diode is in forward bias. There is a flow of charge through the diode, charging up the capacitor. So the voltage at the output can also rise rapidly. However, when the voltage at the input falls, the diode will be reverse biased. The capacitor will only be able to discharge through the resistor. Provided that the time constant of the circuit is long enough, the voltage at the output can only fall slowly. So as the voltage at the input oscillates rapidly up and down, the voltage at the output remains more or less steady. In other words, the d.c. voltage at the output follows the peak value of a.c. signal at the input.

Rectify and filter

The diode detector performs two distinct operations on the a.m. carrier applied to its input.

- It rectifies it, removing the negative part of the a.m. carrier. This adds new components to the frequency spectrum, one of which is the a.f. signal, as shown in Fig 6.19.
- It removes any components above the break frequency of the RC network.

$f_0 = ?$

$R = 100\ \text{k}\Omega$

$C = 100\ \text{pF}$

$$f_0 = \frac{1}{2\pi RC} = \frac{1}{2\pi \times 100 \times 10^3 \times 100 \times 10^{-12}}$$

$$= 1.6 \times 10^4\ \text{Hz or } 16\ \text{kHz}$$

So the a.f. signal below 16 kHz produced by the rectification process appears at the output and the other components above 16 kHz do not.

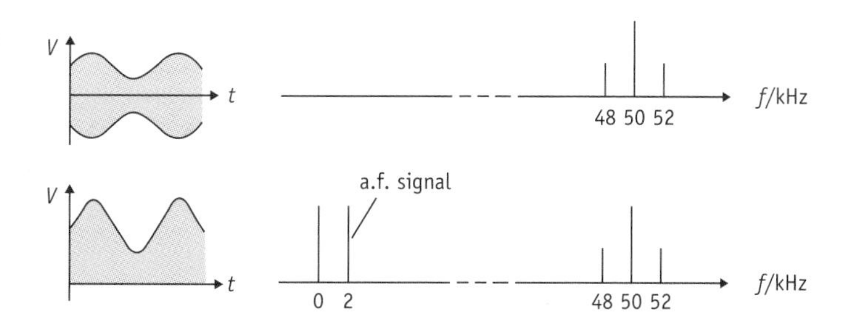

Fig 6.19 Rectification creates a copy of the a.f. signal

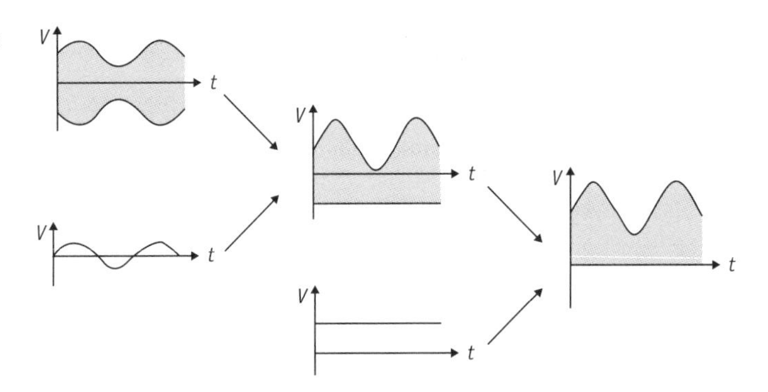

Fig 6.20 Steps in rectifying an a.m. carrier

Creating components

The sequence of graphs in Fig 6.20 shows how rectification can perform the trick of adding extra components to the frequency spectrum of an a.m. carrier. The sequence goes from left to right in two steps.

- Add the right amount of a.f. signal to the a.m. carrier, so that it has a constant lowest value.
- Then add a d.c. signal (at 0 Hz) to shift the whole waveform up. If the right amount is added, the result is a signal which never goes negative, but still oscillates up and down by a varying amount.

Of course, you only need the diode and resistor to do this if the a.m. signal is centred on 0 V, as shown in Fig 6.21. The capacitor in parallel with the resistor makes the RC network into a treble-cut filter, so that only the low frequency components of the rectified signal appear at the output.

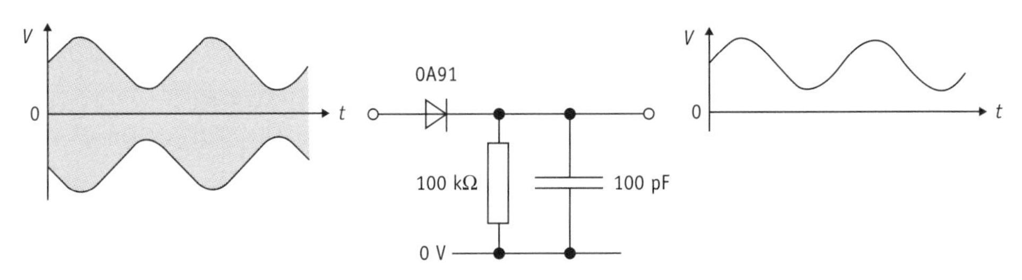

Fig 6.21 Input and output waveforms for a diode detector

6.2 Pulse width modulation

There is one thing that amplitude modulation fails to do. It cannot transmit information about a d.c. signal, one at or near to 0 Hz. This is because the d.c. component at the output of an a.m. demodulator is essentially just a measure of the amplitude of the carrier. So a different modulation technique is required for signals which vary slowly all the way down to 0 Hz.

Pulse coding

Take a careful look at the voltage–time graphs of Fig 6.22.

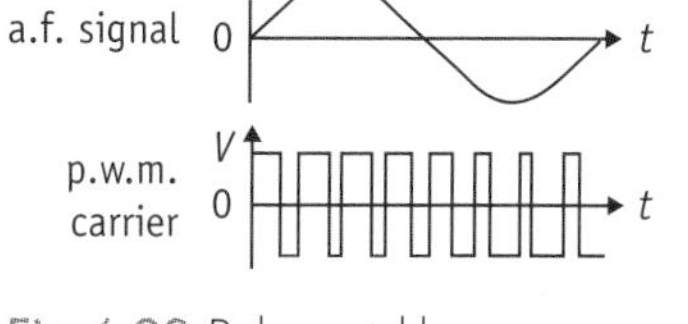

Fig 6.22 Pulse width modulation

The **pulse width modulation** or **p.w.m. carrier** of Fig 6.22 has the following features:

- It is a digital signal, being either positive (the **mark**) or negative (the **space**).
- The time for one cycle (the **period**) remains constant.
- The mark is longer than the space when the a.f. signal is positive.
- The mark is shorter than the space when the a.f. signal is negative.

This means that the moment-by-moment voltage of the a.f. signal is coded as the **mark–space ratio**.

Modulation

In principle, the construction of a p.w.m. modulator is quite straightforward, as shown in Fig 6.23, provided that you know how to construct each block. The graph shows that the output T of the **triangle waveform generator** rise and falls steadily and repeatedly with time. Whenever the signal at T is above the voltage of the a.f. signal A, the output of the **comparator** P is negative, otherwise T is positive. In this way, the mark–space ratio of the p.w.m. carrier is controlled by the a.f. signal, getting bigger and smaller as the voltage rises and falls. Some examples are given in the table.

a.f. signal/V	mark/μs	space/μs	period/μs	mark:space ratio
+5.0	12	4	16	3:1
0.0	8	8	16	1:1
−5.0	4	12	16	1:3

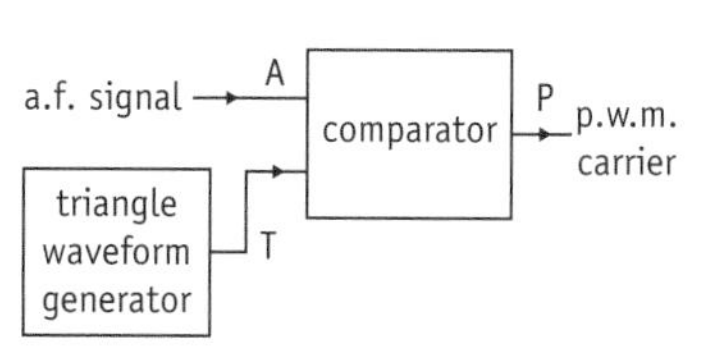

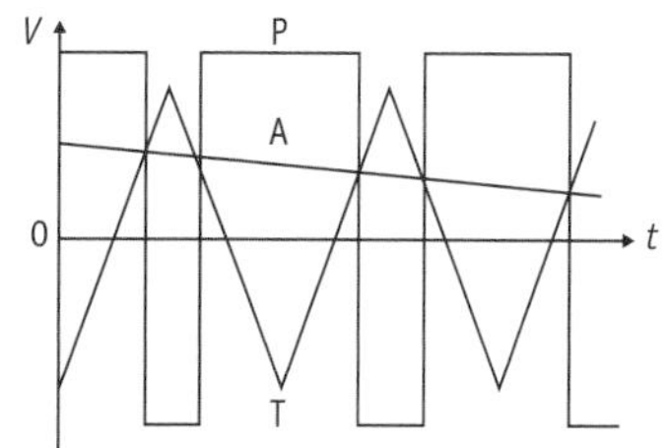

Fig 6.23 Block diagram for a pulse width modulator

Demodulator

Only two processing steps are required to demodulate a p.w.m. carrier when it arrives at the receiver.

- Use a comparator to restore the p.w.m. carrier to its original shape.
- Filter out the high frequency components.

A suitable circuit is shown in Fig 6.24.

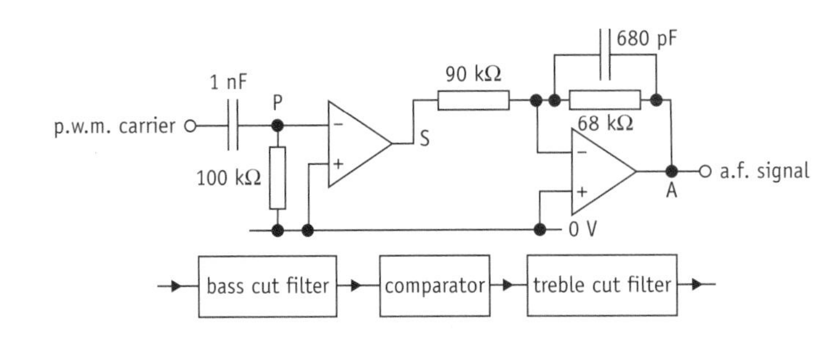

Fig 6.24 A p.w.m. demodulator

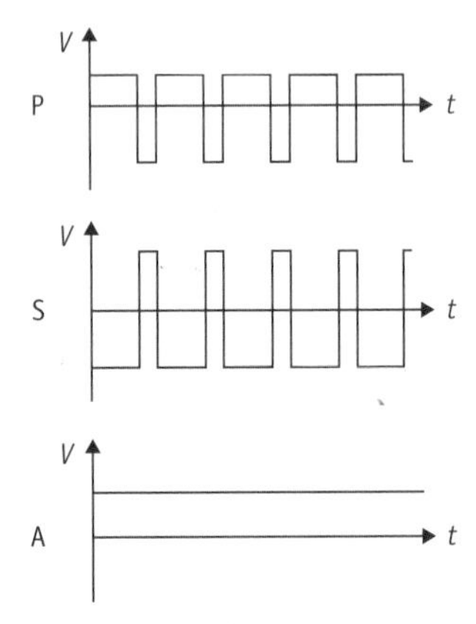

Fig 6.25 Typical waveforms at P, S and A

The waveforms shown in Fig 6.25 will help you to understand how the circuit operates. The p.w.m. carrier at P has a mark–space ratio of 3:1. The op-amp comparator saturates S at either +13 V or −13 V, depending on the current state of the signal at P. So, regardless of what happened to the p.w.m. carrier in transmission, when it arrives at S the mark is fixed at −13 V and the space at +13 V. The average voltage V_{dc} at S can be calculated as follows.

$$V_{dc} = \frac{(-13) \times 3 + (+13) \times 1}{3 + 1} = -6.5\ \text{V}$$

The treble-cut filter has a low frequency gain of −0.76.

$G = ?$

$R_f = 68\ \text{k}\Omega$

$R_{in} = 90\ \text{k}\Omega$

$$G = -\frac{R_f}{R_{in}} = -\frac{68 \times 10^3}{90 \times 10^3} = -0.76$$

The average voltage at the output of the treble-cut filter, is now easily calculated.

$V_{out} = ?$

$V_{in} = -7.5\ \text{V}$

$G = -0.76$

$$V_{out} = GV_{in} = (-0.76) \times -6.5 = +4.9\ \text{V}$$

Providing that the treble-cut filter removes all of the higher frequency components of the signal at S, the output at A will therefore be a steady +5 V while the mark–space ratio of the signal at P remains at 3:1. This means that the demodulator is a good match to the modulator described on the previous page.

Residual carrier signal

The break frequency of the treble-cut filter is set at the top end of the voice message range.

$f_0 = ?$

$R = 68\ \text{k}\Omega$

$C = 680\ \text{pF}$

$$f_0 = \frac{1}{2\pi RC} = \frac{1}{2\pi \times 68 \times 10^3 \times 680 \times 10^{-12}}$$

$$= 3.4\times10^3\ \text{Hz or } 3.4\ \text{kHz}$$

So any information in the a.f. signal between 3.4 kHz and 0 Hz will get through to the output. Ideally, none of the frequency components above 3.4 kHz should get through, but in reality some of them will. This is because the gain of the filter only goes down by ten for every tenfold increase in frequency above the break frequency. The lowest of the high frequency components can be estimated by calculating the **carrier frequency** of the modulated signal, the rate at which the a.f. signal was encoded in the modulator. The period of the p.w.m. signal is 16 μs, giving a carrier frequency of 63 kHz.

$f = ?$

$T = 16\ \mu\text{s}$

$$f = \frac{1}{T} = \frac{1}{16 \times 10^{-6}} = 6.3\times10^4\ \text{Hz or } 63\ \text{kHz}$$

The **residual carrier signal** present at A can be estimated as follows.

- The signal at S swings between +13 V and −13 V, so the amplitude of the component at 63 kHz is likely to be at least 13 V.
- The filter will reduce this component by a factor of 3.4/63 = 0.05.
- The amplitude of the residual carrier signal at 63 kHz is therefore about 0.05 × 13 = 0.7 V.

At first glance, this seems unacceptable. After all, if the signal at A is supposed to be +5.0 V, a rapid variation with an amplitude of 0.7 V is hardly negligible. However, it is actually quite tolerable because you can't hear it on top of the a.f. signal. This is because 63 kHz is well outside the range of human hearing (20 Hz to 20 kHz).

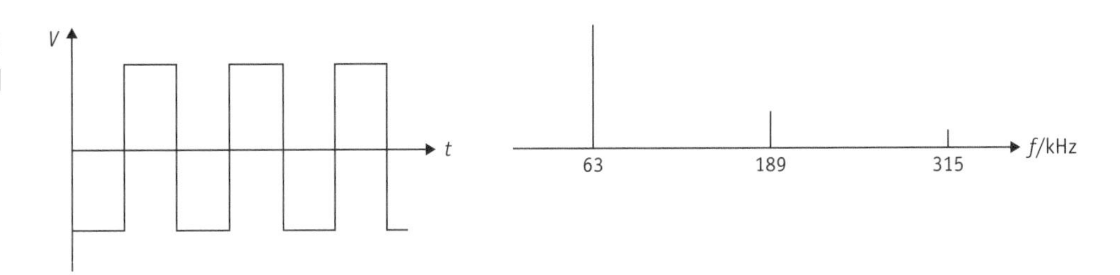

Fig 6.26 Frequency spectrum for a mark–space ratio of 1:1

Frequency spectrum

The sharp transitions of a p.w.m. carrier as it changes between mark and space require a very large bandwidth if they are to survive the passage from modulator to demodulator. However, you only need to send the lower frequency components, as the comparator in the demodulator will **restore** the p.w.m. to its original shape. For example, consider the p.w.m. carrier of Fig 6.26 where the mark–space ratio is 1:1 and the period is 16 μs. Although it is a square wave of frequency 63 kHz, with frequency components at 1×63 kHz, 3×63 kHz, 5×63 kHz . . ., only the lowest frequency component needs to get through to the demodulator, as shown in Fig 6.27.

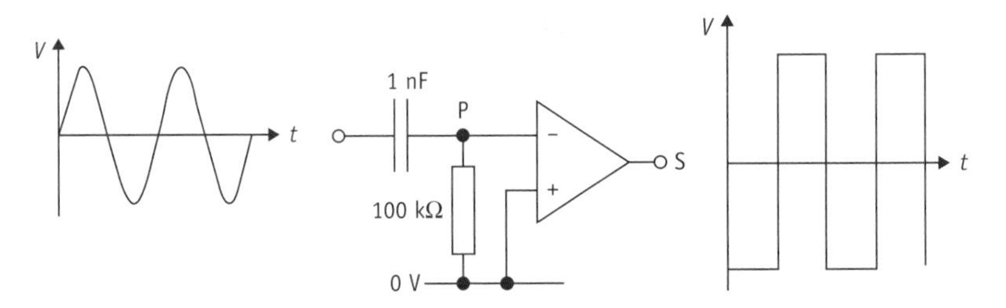

Fig 6.27 Restoring a p.w.m. carrier with a comparator

Matters are more complicated when the mark–space ratio is 1:3, as shown in Fig 6.28. The mark only lasts for 4 μs, with a space of 12 μs. This looks like a square wave with a period of 8 μs where every other pulse has been suppressed, suggesting that it can be approximated by a sine wave at 126 kHz strongly amplitude modulated by another one at 63 kHz, giving the frequency spectrum shown. A little thought should convince you that the spectrum is the same when the mark–space ratio is 3:1.

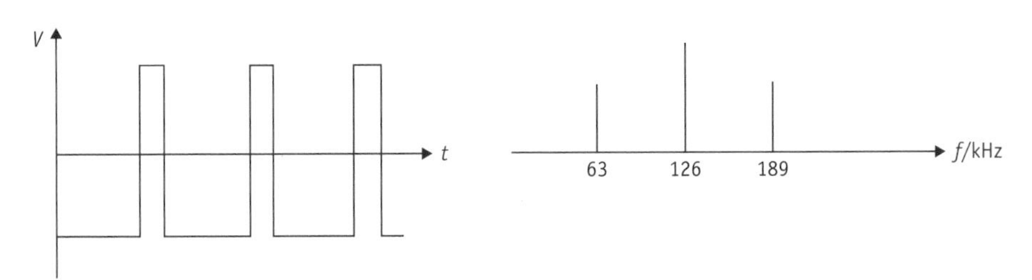

Fig 6.28 Approximate frequency spectrum for a mark–space ratio of 1:3

So if the mark–space ratio is kept within the limits of 3:1 and 1:3, you can just about get away with transmitting only the frequencies between 63 kHz and 189 kHz. This leads to the following useful rule for the bandwidth Δf_{pwm} of a p.w.m. carrier with a carrier frequency of f_c.

$$\Delta f_{pwm} = 2f_c$$

Of course, if the mark–space ratio goes outside the range of 1:3 to 3:1, you will need a larger bandwidth.

Sampling rate

On the previous page you found out how the bandwidth Δf_{pwm} needed for a p.w.m. carrier depends on the carrier frequency f_c, the frequency of the triangle waveform generator in the modulator. As you will find out in the next chapter, it is often important to keep the bandwidth to a minimum. For a p.w.m. system, this clearly means having a low value for the carrier frequency. However, there is a limit to how low you can make the carrier frequency without severely distorting the a.f. signal.

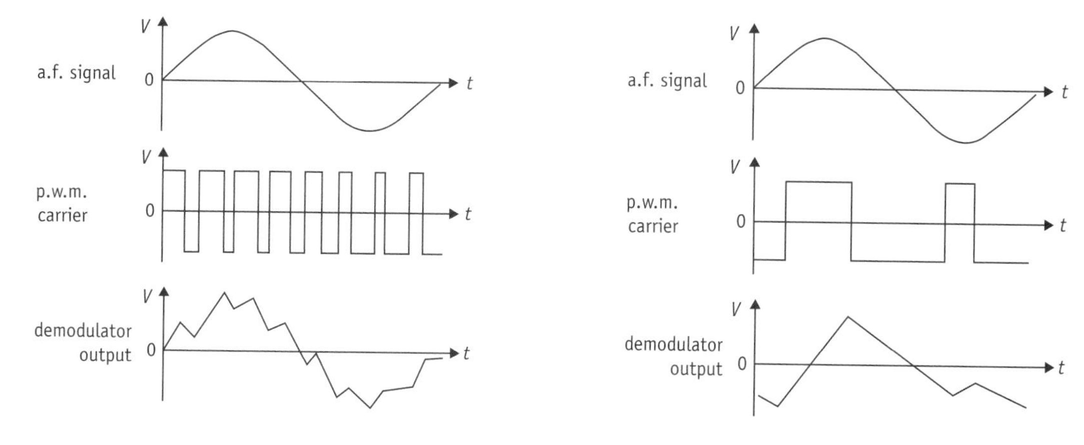

Fig 6.29 The a.f. signal is sampled eight times in each cycle

Fig 6.30 Sampling only twice in each cycle

The a.f. signal shown in Fig 6.29 is **sampled** eight times in each cycle. This means that the p.w.m. carrier contains eight sets of information about each cycle of the a.f. signal, enough for the demodulator to reconstruct the approximate copy shown. Notice how the residual carrier signal masks the correct shape of the a.f. signal. This gets worse as the **sampling rate** is reduced. In fact, once you get down to sampling just twice in each cycle (Fig 6.30), only just enough information gets passed on to the demodulator to reconstruct a signal with the same frequency and amplitude as the original. This is known as the **Nyquist criterion**: the sample rate must be at least twice the signal frequency.

It clearly pays to have a high sampling rate if you want a reasonably accurate copy of the a.f. signal at the output, and a residual carrier well beyond the range of human hearing. But increasing f_c increases the bandwidth, which is never a good thing. If you compromise on at least 8 samples per cycle, then $f_c = 8f_s$ and the bandwidth can be estimated with this formula.

$$\Delta f_{pwm} = 16f_s$$

So what is the bandwidth required for a voice message in the frequency range 300 Hz to 3.4 kHz transmitted by pulse width modulation?

$\Delta f_{pwm} = ?$ $\qquad$ $\Delta f_{pwm} = 16f_s = 16 \times 3.4\times10^3 = 5.4\times10^4$ Hz or 54 kHz

$f_s = 3.4$ kHz

This is much bigger than the 7 kHz required for amplitude modulation.

Triangle waveforms

The triangle waveform generator shown in Fig 6.31 has two distinct blocks.

- A **ramp generator** whose output rises or falls steadily.
- A **Schmitt trigger** whose output saturates high or low.

The blocks are connected such that the output of one provides the input signal for the other. Why this results in continual oscillation should become clear after you have found out how each block behaves on its own.

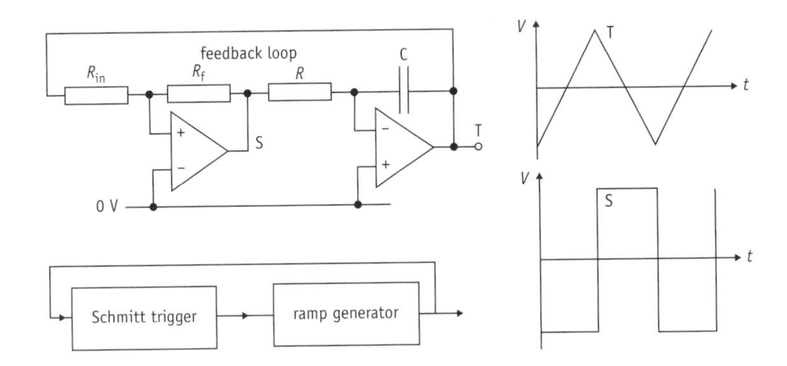

Fig 6.31 Triangle waveform generator

Ramp generator

If the circuit of Fig 6.31 is to be the triangle waveform generator for the p.w.m. modulator of Fig 6.22 on page 101, the output of the ramp generator must rise from −10 V to +10 V and back to −10 V in 16 μs.
The time constant RC of the ramp generator can be calculated with this formula.

$$\Delta V_{out} = -\frac{V_{in}\Delta t}{RC}$$

ΔV_{out} is the change of voltage at the output in the time interval Δt when the voltage at the input is V_{in}. Since the ramp generator input S is connected to the output of a saturated op-amp, you can safely assume that S will be either at +13 V or −13 V. Let's set T to −13 V for the calculation of the time constant.

$RC = ?$

$\Delta V_{out} = +10 - (-10)$
$= +20$ V

$V_{in} = -13$ V

$\Delta t = \frac{1}{2}16\mu s = 8\mu s$

$$\Delta V_{out} = -\frac{V_{in}\Delta t}{RC}$$

$$RC = -\frac{V_{in}\Delta t}{\Delta V_{out}} = -\frac{-13 \times 8 \times 10^{-6}}{20} = 5.2\times10^{-6}\text{ s}$$

Now choose a value for C (such as 1.5 nF) and calculate the value of R.

$R = ?$

$C = 1.5$ nF

$RC = 5.2\times10^{-6}$ s

$$R = \frac{5.2 \times 10^{-6}}{1.5 \times 10^{-9}} = 3.4\times10^{3}\ \Omega \text{ or } 3.4\text{ k}\Omega$$

So if we set R to 3.3 kΩ (the nearest obtainable value), the ramp generator should behave as specified.

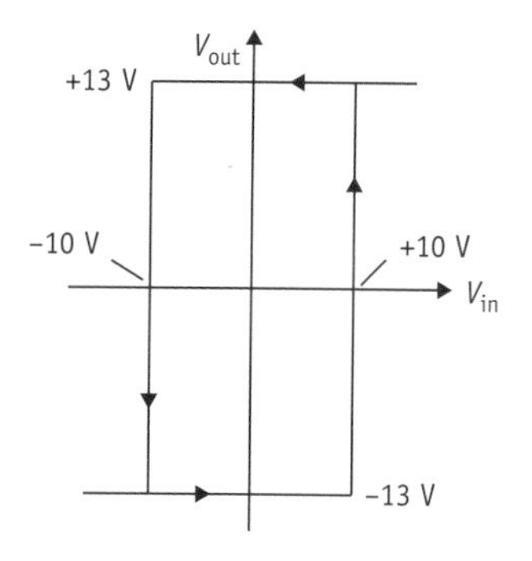

Fig 6.32 Schmitt trigger transfer characteristic

Schmitt trigger

The Schmitt trigger is required to have the transfer characteristic shown in Fig 6.32, with trip points at +10 V and −10 V. This means that the reaction of the op-amp to signals at its input T depends on the current state of its output S.

- When S is +13 V, T has to go below −10 V to change S to −13 V.
- When S is at −13 V, T has to go above +10 V to change S to +13 V.

The trip points are set by the values of the feedback resistor R_f and the input resistor R_{in}. Start off by assuming that S is saturated at −13 V and that T is sourcing current I into the input resistor, as shown in Fig 6.33.

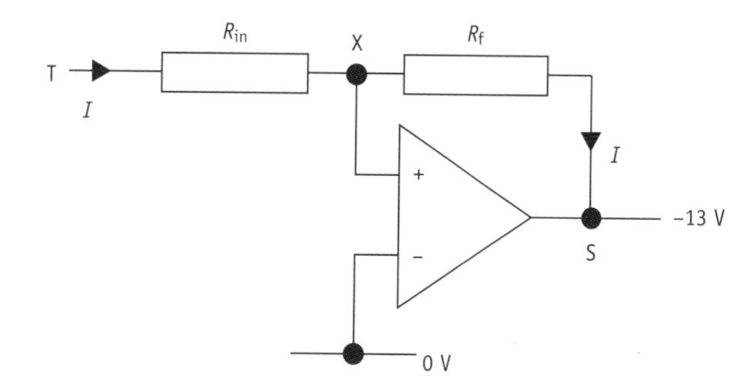

Fig 6.33 S changes state when X goes above 0 V

Provided that point X is below 0 V, the op-amp output S will remain at −13 V, However, if T is high enough, then X will go above 0 V, resulting in an immediate change of voltage at S from −13 V to +13 V. So if T is at +10 V, then X must be almost at 0 V when S is −13 V. If you choose a convenient value for R_f, such as 39 kΩ, then you can calculate the current I in it.

$I = ?$

$V = 0 - (-13) = +13 \text{ V}$

$R = 39 \text{ k}\Omega$

$$I = \frac{V}{R} = \frac{13}{39 \times 10^3} = 3.3 \times 10^{-4} \text{ A}$$

Since the current in the op-amp input is negligible, there will be the same current in both of the resistors, allowing you to calculate the value of R_{in}.

$R = ?$

$V = +10 - 0 = 10 \text{ V}$

$I = 3.3 \times 10^{-4} \text{ A}$

$$R = \frac{V}{I} = \frac{10}{3.3 \times 10^{-4}} = 3.0 \times 10^4 \ \Omega \text{ or } 30 \text{ k}\Omega$$

Positive feedback

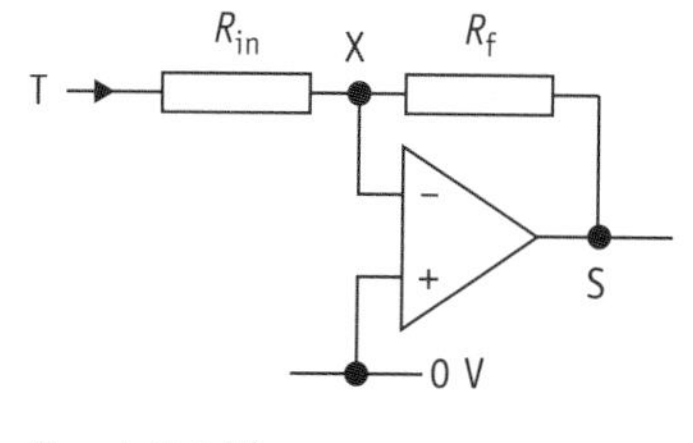

Fig 6.34 This is an inverting amplifier, not a Schmitt trigger

It is easy to confuse the circuit of Fig 6.33 with that of an inverting amplifier. Please don't. They have completely different behaviour. The **positive feedback** employed in a Schmitt trigger means that it has only two possible stable states, saturated at either +13 V or −13 V. An inverting amplifier (Fig 6.34) employs **negative feedback**, forcing it to set its output at whatever value results in 0 V at the virtual earth of point X.

6.3 Frequency modulation

High-quality radio broadcasts use **frequency modulation** (or **f.m.**) to encode speech or music signals. It is more complicated to achieve than amplitude modulation and requires a much larger bandwidth, but in return it offers the advantage of a much greater immunity from noise and interference. The idea behind f.m. is very simple. Make the frequency of the h.f. carrier follow the moment-by-moment voltage of the a.f. signal, as shown in Fig 6.35, so that the information is coded by the period of each cycle rather than its amplitude.

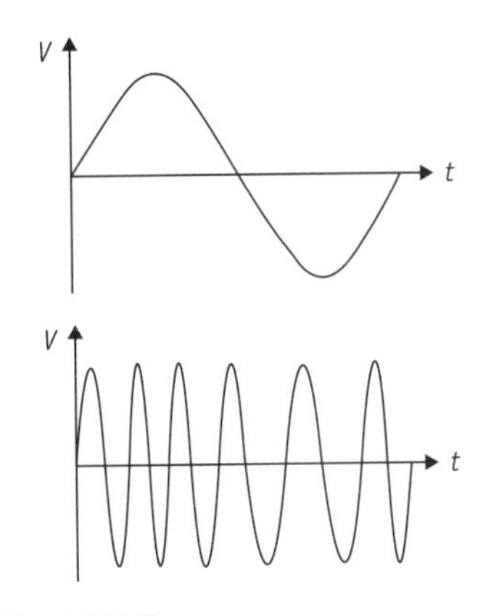

Fig 6.35 Frequency modulation

Modulator

Fig 6.36 shows the block and circuit diagram of an f.m. modulator. It is essentially a voltage-to-frequency converter.

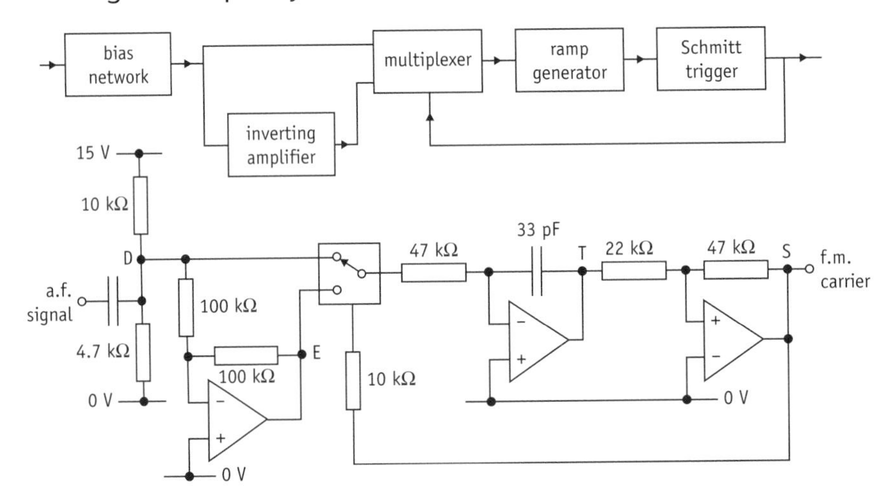

Fig 6.36 Block and circuit diagrams for an f.m. modulator

Here is the function of each block, working from left to right.

- The bias network adds a steady +5 V to the a.f. signal present at the input.
- The inverting amplifier multiplies the signal at D by −1 and places it at E.
- The multiplexer lets either of the signals at D or E through to the ramp generator, depending on the signal fed back from the Schmitt trigger.
- The ramp generator output rises or falls steadily at a rate determined by the voltage at D.
- The Schmitt trigger changes state each time its input goes above +6 V or below −6 V.

The end result is a square wave at S whose frequency depends on the voltage at D. Increasing the voltage at D has the following effect on the circuit.

- Increases the rate at which the voltage at T rises or falls.
- Decreases the time taken for T to move from one trip point to another.
- Decreases the period of the square wave at S.

So a momentary increase voltage at D caused by the a.f. signal at the input results in an increase in frequency of the f.m. carrier at the output.

Carrier frequency

What is the frequency of the h.f. carrier produced by the circuit of Fig 6.36? Start off by finding the current in the bias network.

$I = ?$

$V = 15\text{ V}$

$R = 10 + 4.7 = 14.7\text{ k}\Omega$

$$I = \frac{V}{R} = \frac{15}{14.7 \times 10^3} = 1.0 \times 10^{-3}\text{A}$$

Then calculate the voltage drop across the 4.7 kΩ resistor to find the voltage at D.

$V = ?$

$I = 1.0\times10^{-3}\text{ A}$

$R = 4.7\text{ k}\Omega$

$$V = IR = 1.0\times10^{-3} \times 4.7\times10^3 = 4.7\text{ V}$$

Now calculate the rate at which the output T of the ramp generator rises when the multiplexer selects the −4.7 V signal at E.

$\frac{\Delta V_{out}}{\Delta t} = ?$

$V_{in} = -4.7\text{ V}$

$RC = 47\times10^3 \times 33\times10^{-12} = 1.6\times10^{-6}\text{ s}$

$$\frac{\Delta V_{out}}{\Delta t} = -\frac{V_{in}}{RC} = -\frac{-4.7}{1.6 \times 10^{-6}} = 3.0 \times 10^6\text{ V s}^{-1}$$

So if the output is rising at $3.0\times10^6\text{ V s}^{-1}$, how long does it take to get from one trip point of the Schmitt trigger to the other? The trip points of the Schmitt trigger are determined by the ratio of the input and output resistors. Assume that the output S is saturated at −13 V and calculate the voltage at the input necessary to hold the non-inverting input of the op-amp at 0 V.

$V_{in} = ?$

$V_{out} = -13\text{ V}$

$R_f = 47\text{ k}\Omega$

$R_{in} = 22\text{ k}\Omega$

$$I = \frac{V}{R} = \frac{13}{47 \times 10^3} = 2.8 \times 10^{-4}\text{A}$$

$$V = IR = 2.8\times10^{-4} \times 22\times10^3 = 6.1\text{ V}$$

T rises at $3.0\times10^6\text{ V s}^{-1}$. How long does it take to get from − 6.1 V to +6.1 V?

$\Delta t = ?$

$\Delta V = +6.1 - (-6.1) = 12.2\text{ V}$

$\frac{\Delta V}{\Delta t} = 3.0 \times 10^6\text{ V s}^{-1}$

$$\frac{\Delta V}{\Delta t} = \frac{12.2}{\Delta t} = 3.0 \times 10^6$$

$$\Delta t = \frac{12.2}{3.0 \times 10^6} = 4.1 \times 10^{-6}\text{s}$$

Of course, in each cycle of T it rises by 12.2. V and then falls again by 12.2 V, requiring a time of 8.2×10^{-6} s in all. You are now in a position to calculate the frequency of the signal at S.

$f = ?$

$T = 8.2\times10^{-6}\text{ s}$

$$f = \frac{1}{T} = \frac{1}{8.2 \times 10^{-6}} = 1.2 \times 10^5\text{Hz or 120 kHz}$$

It is not difficult to run through the calculation again to show that raising the voltage at D by 1.0 V increases the frequency at S by about 25 kHz.

Demodulator

The f.m. demodulator of Fig 6.37 is basically a frequency-to-voltage converter.

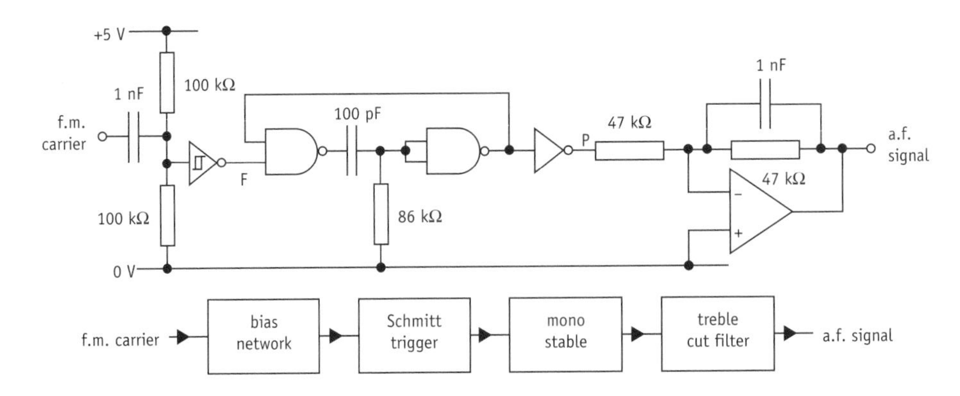

Fig 6.37 Circuit and block diagrams for an f.m. demodulator

The block diagram shows that the circuit has four distinct sub-systems.

- A bias network to centre the incoming f.m. carrier on 2.5 V, halfway through the region of uncertainty between logic 1 (5 V to 3 V) and logic 0 (2 V to 0 V).
- A Schmitt trigger to sharpen up the rising and falling edges of the f.m. carrier, as illustrated in Fig 6.38. This converts the analogue f.m. carrier into a digital signal. Notice that the Schmitt trigger loses all information about the amplitude of the f.m. carrier, only passing on its frequency.

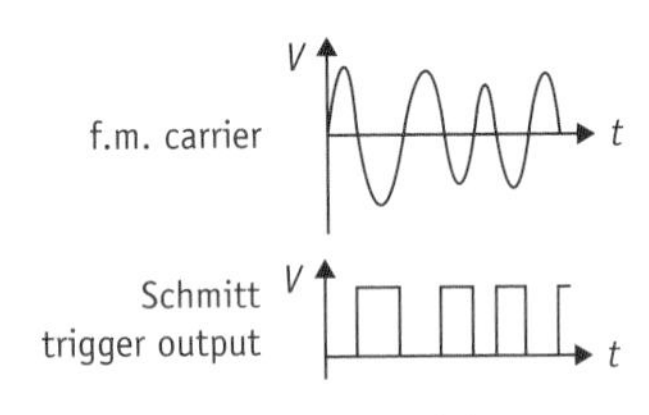

Fig 6.38 The effect of the Schmitt trigger

- A monostable to produce one 5 V pulse of fixed duration for each cycle of the f.m. carrier, as shown in Fig 6.39. Increasing the frequency of the carrier brings these pulses closer together, increasing the average voltage of the signal at P.
- A treble-cut filter to remove the high frequency components of the pulses at P. The remaining low frequency component is the a.f. signal and a d.c. component. The latter is easily removed by a passive bass-cut filter (not shown in Fig 6.37).

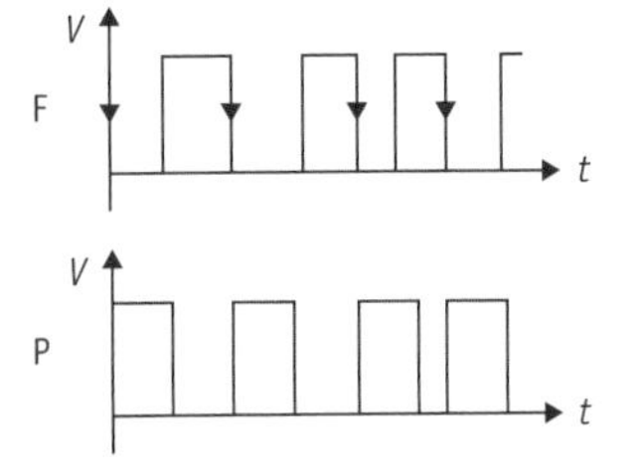

Fig 6.39 The effect of the monostable

The duration of the pulses at P and the break frequency of the treble-cut filter both have to be carefully matched to the properties of the modulator which produced the f.m. carrier in the first place.

Monostable

Suppose that the modulator is that of Fig 6.36. It has the following properties.

- An unmodulated h.f. carrier at 120 kHz.
- A maximum **deviation frequency** of 20 kHz, limiting the f.m. carrier to between 100 kHz and 140 kHz.

This means that the period of the signal at F can be anywhere between 10 μs and 7 μs. The pulse duration at P needs to be a bit less than the lower of these values, so that the pulse at P has always finished before the next falling edge arrives at F.

$T = ?$ $\quad T = 0.7RC = 0.7 \times 86{\times}10^3 \times 100{\times}10^{-12}$

$R = 86\ \text{k}\Omega$ $\quad T = 6.0{\times}10^{-6}$ s or 6 μs

$C = 100$ pF

You will recall from your work at AS that the output of a monostable pulses low each time a falling edge enters it. The NOT gate in Fig 6.37 converts that low pulse into a high one. This means that increasing the frequency of the f.m. carrier increases the average voltage entering the filter, rather than decreasing it.

Filter

Fig 6.40 shows the signals arriving at the treble-cut filter when the f.m. carrier is at its highest and lowest values.

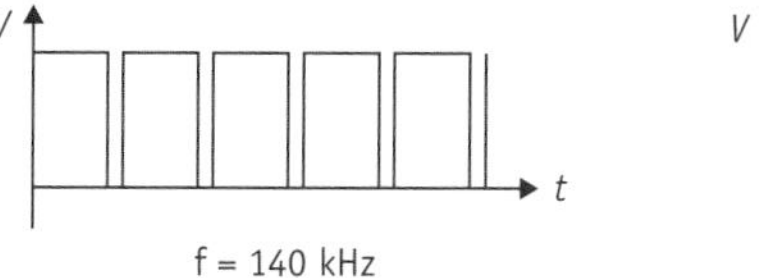

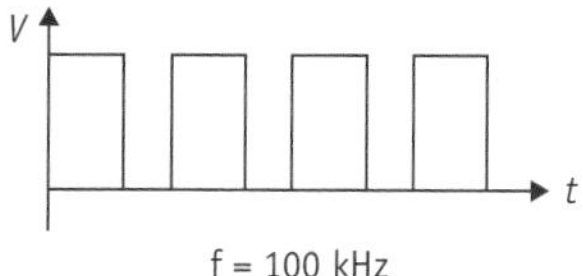

Fig 6.40 Signals at P for maximum and minimum frequencies of the f.m. carrier

The average voltage at P, when the period is 7 μs, can be calculated as follows.

$$V_{avge} = \frac{5 \times 6 + 0 \times 1}{7} = 4.3\ \text{V}$$

You can check for yourself that when P is 10 μs, V_{avge} drops to 3.0 V. So the maximum amplitude of the a.f. signal at the output of the filter will be $(4.3 - 3.0)/2 = 0.65$ V. The break frequency matches the voice message range.

$f_0 = ?$ $\quad f_0 = \dfrac{1}{2\pi RC} = \dfrac{1}{2\pi \times 47 \times 10^3 \times 1 \times 10^{-9}}$

$R = 47\ \text{k}\Omega$ $\quad = 3.4 \times 10^3$ Hz or 3.4 kHz

$C = 1$ nF

How much residual carrier will there be at the output of the filter? The lowest component of the h.f. carrier at P will have an amplitude of about 2.5 V and frequency of about 120 kHz. The residual carrier signal is therefore likely to have an amplitude of $2.5 \times (3.4/120) = 0.07$ V. Although this is about 10% of the maximum amplitude of the a.f. signal, you shouldn't worry about it as it is well outside the range of human hearing!

Bandwidth

So what is the bandwidth required to transmit an f.m. carrier? It is clearly going to depend on two separate factors.

- The maximum deviation frequency f_{dev}.
- The maximum signal frequency f_s.

The first factor is obvious, because it sets the upper and lower limits on the frequency of the h.f. carrier $f_{carrier}$ as it is modulated.

$$f_{upper} = f_{carrier} + f_{dev}$$

$$f_{lower} = f_{carrier} - f_{dev}$$

The second factor is more subtle. When it is modulated, the h.f. carrier never spends very long at any particular value. This introduces a number of other sidebands on either side of the h.f. carrier, separated from it according to the value of the a.f. signal frequency. So the bandwidth Δf_{fm} of an f.m. carrier is given by this approximation.

$$\Delta f_{fm} \simeq 2(f_s + f_{dev})$$

As you will see below, the bandwidth can be optimised by setting f_{dev} to $1.5f_s$, giving this useful rule for the bandwidth.

$$\Delta f_{fm} = 5f_s$$

So what is the bandwidth required to send voice messages with the usual frequency range of 300 Hz to 3.4 kHz by frequency modulation?

$\Delta f_{fm} = ?$ $\qquad$ $\Delta f_{fm} = 5f_s = 5 \times 3.4 \times 10^3 = 1.7 \times 10^4$ Hz or 17 kHz

$f_s = 3.4$ kHz

This is quite a bit more than the 7 kHz which is required for amplitude modulation but a lot less than the 54 kHz required for pulse width modulation.

Frequency spectrum

The voltage–time graphs on the next page attempt to show how the extra sidebands due to f_s arise. They are for the following situation where the a.f. signal has its maximum amplitude and frequency.

- The unmodulated h.f. carrier is at 26 kHz.
- The a.f. signal makes the h.f. carrier alternate between 20 kHz and 32 kHz, so the maximum deviation frequency is 6 kHz.
- The frequency of the a.f. signal is 4 kHz.

Now although the modulation process will smoothly move the h.f. carrier from 20 kHz to 32 kHz and back again, this is approximately the same as alternating the h.f. carrier between the two values. In other words, the 20 kHz carrier is switched on and off, as shown in Fig 6.41.

The 20 kHz carrier has been amplitude modulated by the a.f. signal, so that its amplitude rises and falls as time goes on. The frequency spectrum of Fig 6.41 shows the two sidebands at 20 + 4 = 24 kHz and 20 − 4 = 16 kHz required for this modulation.

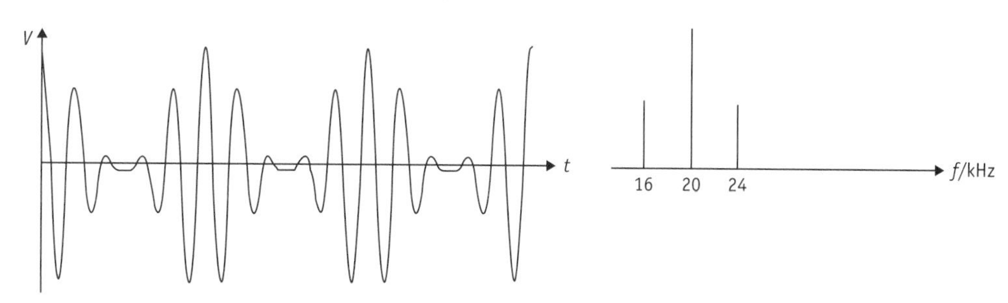

Fig 6.41 Switching the 20 kHz carrier on and off

The 32 kHz carrier must be switched on when the 20 kHz is switched off. This is shown in Fig 6.42. This time, the sidebands are at 28 kHz and 36 kHz. Notice that, to get a depth of modulation which reduces the carrier amplitude to zero, each sideband needs to be half the size of the carrier.

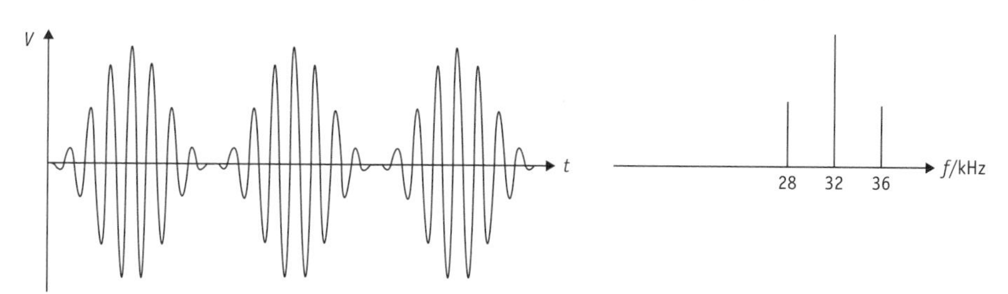

Fig 6.42 Switching the 32 kHz carrier off and on

Fig 6.43 shows what happens when all six frequency components are added together. As you can see, the result is a signal whose frequency alternates between two different values. Notice that, by setting $f_{dev} = 1.5f_s$, the spikes are evenly spread across the frequency spectrum, making efficient use of the bandwidth.

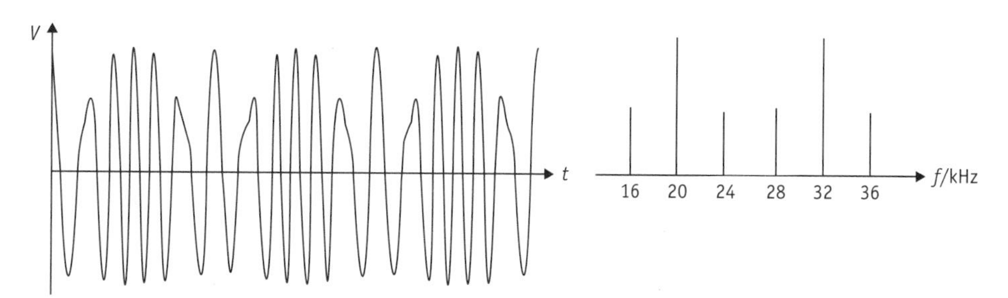

Fig 6.43 The full spectrum

6.4 Signal transmission

A modulated carrier can be transmitted from one place to another in three different ways.

- Along wires as electric current.
- Through the atmosphere as radio waves.
- Through optical fibres as infrared.

Each method of transmission has advantages and disadvantages compared with the other two, but they all suffer from the same problem. The modulated carrier that leaves the **transmitter** is never the same as the modulated carrier that arrives at the **receiver**. Three things happen to the modulated carrier.

- Its amplitude decreases as it travels away from the transmitter.
- **Noise** from the transmission medium is added.
- Other electrical devices add **interference**.

This section of the chapter will explain this in more detail and show how some of these problems can be overcome by selecting an appropriate scheme of modulation.

Noise and interference

All electronic components, including the wires, generate **noise**. As you can see form Fig 6.44, this is a random signal, covering the whole frequency spectrum, and is usually quite small. It arises from random fluctuations to the distribution of charge in conducting materials. The only way of getting rid of it completely is to

Fig 6.44 Noise and interference compared

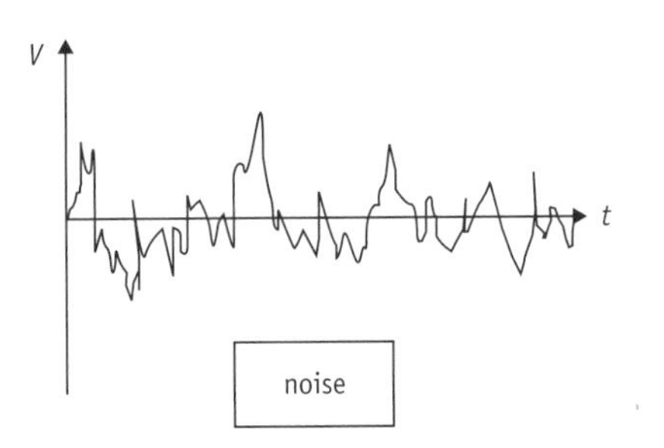

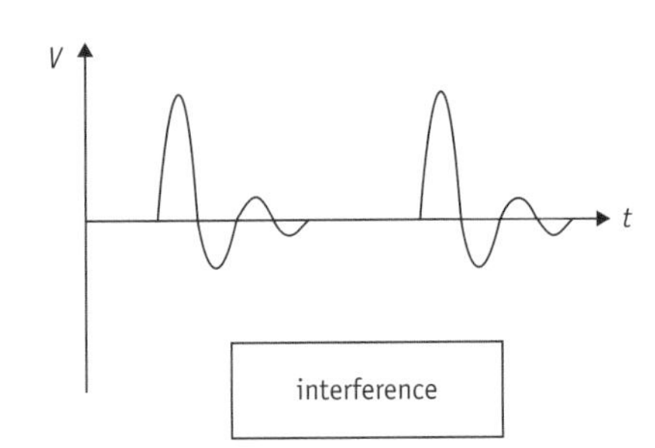

cool the material down to −273°C, absolute zero!

Interference usually looks quite different from noise, often carrying information about its source. It is caused by electrical signals outside the transmission system, so often has some regularity to its shape, as shown in Fig 6.44. Here are some causes of interference.

- The capacitance between adjacent wires allows signals to stray from one wire to the other.
- Radio waves emitted by electric sparks can be picked up by nearby circuits.
- Computers and other devices which have high frequency oscillators emit signals which can look like modulated carriers.

As you will find out later, various techniques can be used to reduce interference to a manageable level, depending on the transmission method being used.

Twisted-pair cable

Interference is almost inevitable, given the large number of electrical and electronic devices in our environment. So a great deal of design effort goes into reducing its effect on communication systems. For example, Fig 6.45 shows one useful technique for reducing interference in systems that send analogue signals through long stretches of wire.

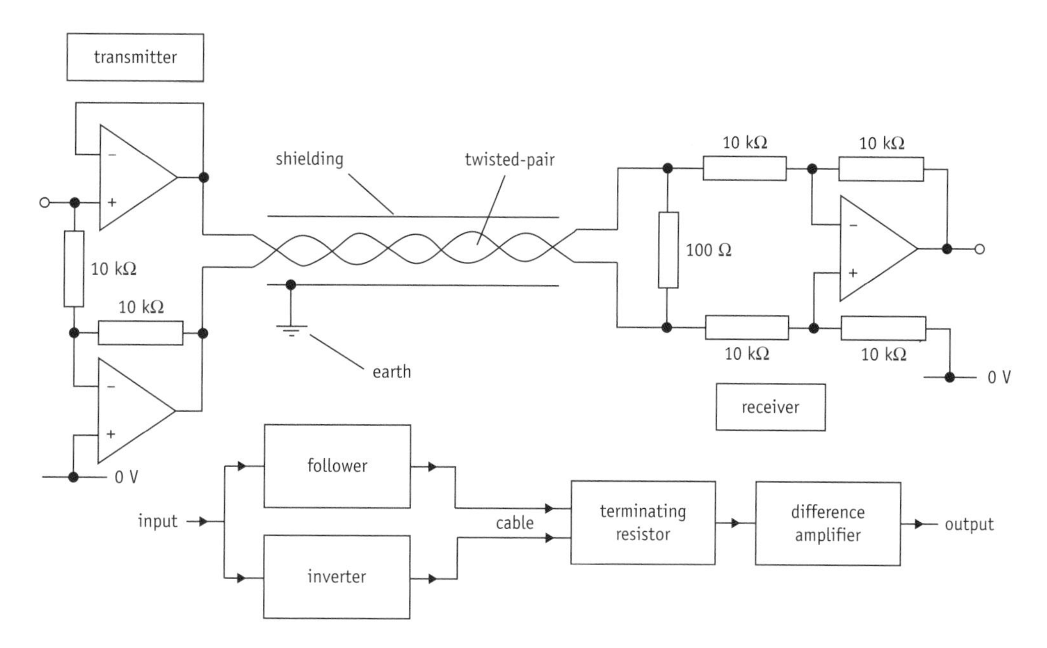

Fig 6.45 Shielded twisted-pair cable eliminates a lot of interference

The modulated carrier (or **signal**) is sent from transmitter to receiver through a pair of wires wrapped around each other. One wire carries the signal directly, the other carries an inverted copy. Since the two wires follow the same path, they are likely to acquire the same interference. So the two wires in the **twisted-pair cable** have the same interference but opposite signals. These appear across the 100 Ω resistor which terminates the cable at the receiver. The interference makes both ends of the resistor have the same voltage, so is ignored by the difference amplifier. The signal, however, makes the ends of the resistor have opposite voltages, so these are passed on to the output of the receiver.

Shielding

Fig 6.45 shows an earthed conductor wrapped around the twisted-pair cable. This **shielding** acts as further protection against interference because radio waves penetrate it with difficulty. Shielding is also useful because it stops the alternating currents in the twisted-pair cable from emitting radio waves which could, in turn, cause interference to other systems. This is already going to be at a low level because the current in one wire is always in the opposite direction to the current in the other wire, cancelling out their effects.

Attenuation

The amplitude of a modulated carrier decreases as it passes along a cable. It is **attenuated**, as shown in the graph of Fig 6.46.

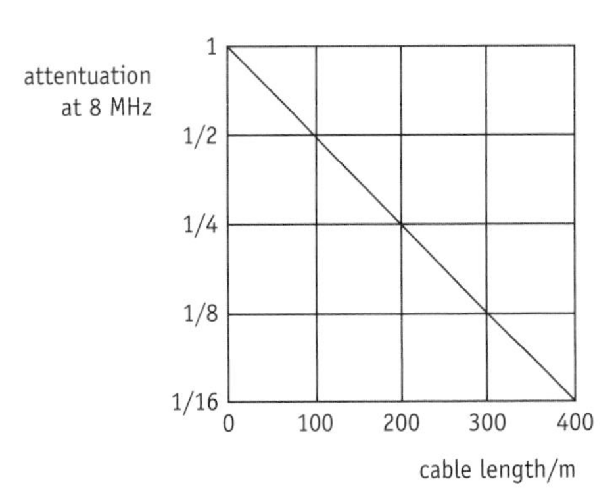

Fig 6.46 Typical attenuation for twisted-pair cable

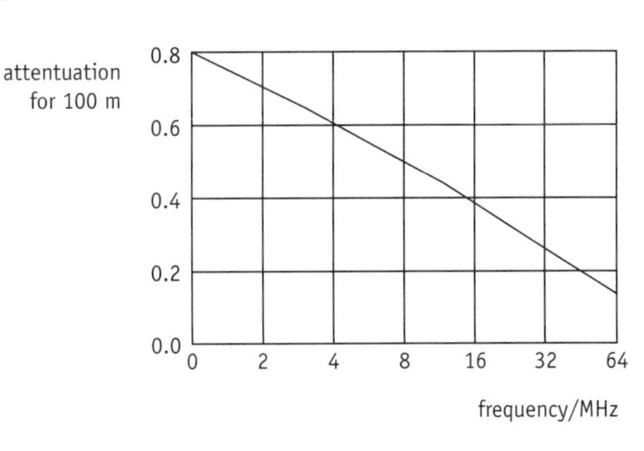

Fig 6.47 Attenuation gets worse as frequency increases

Attenuation is a consequence of the finite resistance of the cable. Alternating current in a resistor generates heat, so the energy of a signal gradually dissipates as it passes down a cable. As you can see from Fig 6.46, the amplitude of a signal at 8 MHz is typically halved for every 100 m of twisted-pair cable that it passes through. To make matters worse, the degree of attenuation increases with increasing frequency of the signal (Fig 6.47). So at 64 MHz, almost 90% of the signal is lost in every 100 m of cable.

Signal-to-noise ratio

Fig 6.48 on the next page shows the fate of an a.m. carrier as it travels down a cable. The top graph shows the a.m. carrier (the **signal**) as it enters the cable. The noise in the cable is comparatively small at this stage, hardly affecting the signal, so the **signal-to-noise ratio** (or **SNR**) is much greater than 1. As the signal passes along the cable, two things happen.

- The signal decreases in amplitude as it is attenuated.
- The amount of noise and interference increases.

The middle graph shows the situation when the signal has travelled far enough for the attenuation to reduce its amplitude enough to make the SNR about 1. At this point, the signal and noise/interference have the same size, making it difficult to extract information from the modulated carrier. The bottom graph shows what happens when the signal is allowed to travel so far down the cable that the SNR becomes less than 1. The signal has been obliterated by the noise, and any information that it contains will be almost impossible to extract at the receiver.

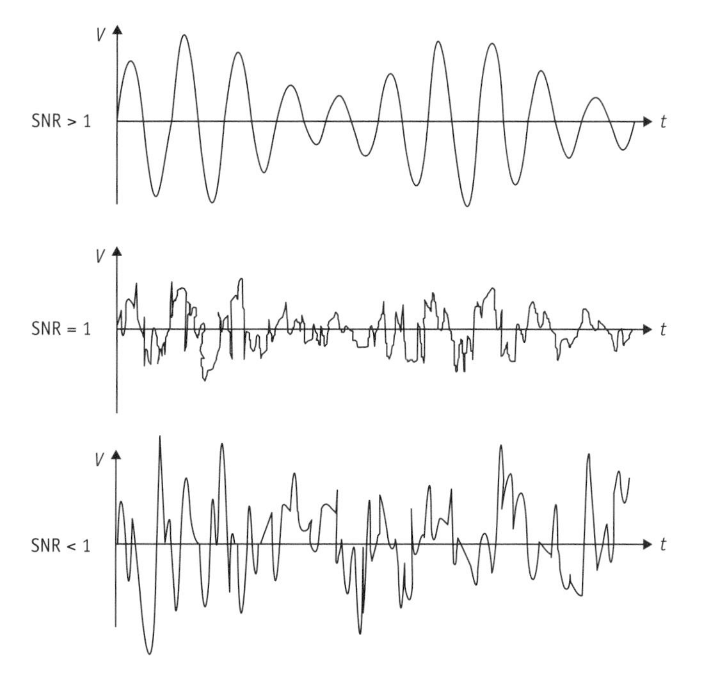

Fig 6.48 The signal-to-noise ratio decreases as the a.m. carrier travels down the cable

Optical fibre

Fig 6.49 shows a transmission system which uses optical fibre as the link between transmitter and receiver. The LED sends pulses of **infrared** light down the fibre, a length of very transparent glass. When the pulses leave the far end of the fibre, the **phototransistor** sinks pulses of current in the pull-up resistor.

Optical fibre has several advantages compared with twisted-wire cable.

- It has almost total immunity from interference. This is because the signal is carried by infrared light, not electric current, and stray light signals are blocked by the black plastic lining of the fibre.
- It has a very low attenuation, with signals typically halving in amplitude for every kilometre of fibre.
- It is a very low noise transmission medium.
- It has a comparatively high bandwidth, equivalent to 50 GHz for cable.

For these reasons, optical fibre is the medium of choice for long-distance high-quality communication. Although it is possible to use optical fibre to transmit analogue signals such as a.m. carriers, it is more often used for digital signals, such as p.w.m. and f.m. carriers.

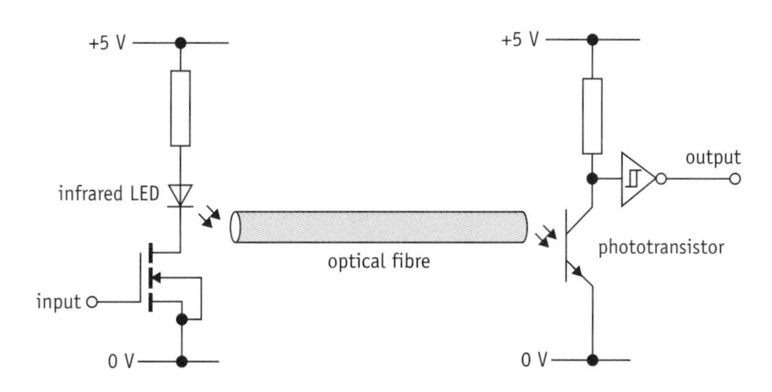

Fig 6.49 Optical fibre transmission system

Radio waves

Radio waves are essentially electrical signals which have escaped the confines of their metal conductors and are free to travel through empty space. They are used for broadcast TV and radio, mobile phones and Wi-Fi access for computers. Fig 6.50 shows details of the circuitry required for communication by radio waves.

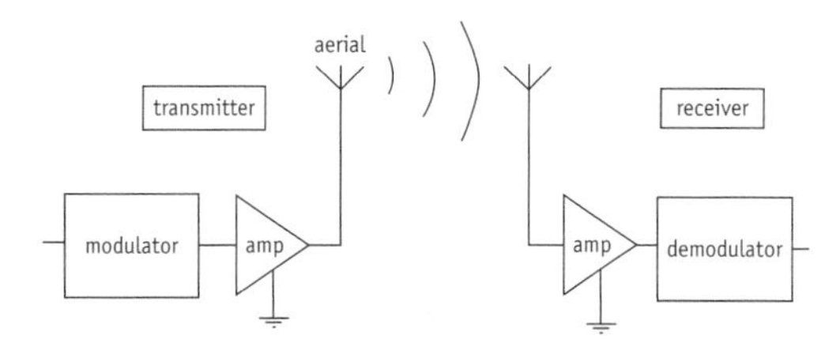

Fig 6.50 Radio transmission system

This is how the system operates (more detail will be given in the next chapter).

- An amplifier in the transmitter applies a modulated carrier to the aerial.
- Alternating current in the **transmitter aerial** creates radio waves.
- The waves spread out in all directions at the speed of light.
- Waves are absorbed by the **receiver aerial**, setting up an alternating current.
- The receiver amplifier converts the current at its input into a modulated carrier.

The aerials are just pieces of wire whose length should ideally be carefully matched to the frequency of the h.f. carrier. It also helps if the 0 V rails of both amplifiers are earthed.

Eliminating noise

Radio wave communication has two disadvantages.

- The signal amplitude drops rapidly as the waves spread out from the transmitter aerial, so the signal-to-noise ratio in the receiver is often not good.
- There is no way of stopping interference from the many other emitters of radio waves in our environment.

Fortunately, it usually possible to completely restore modulated f.m. and p.w.m. carriers in the receiver. Take a look at Fig 6.51. The f.m. modulated signal on the left has acquired a lot of noise and interference on its way from the transmitter. This affects mostly the amplitude of the signal, information which is ignored by the Schmitt trigger. So unless the SNR is particularly poor, the output of the Schmitt trigger is a faithful copy of the modulated carrier in the transmitter. Unfortunately, this technique cannot be used to separate out noise and interference from an a.m. carrier. Radio broadcasts on the MW band use amplitude modulation, so they often suffer from background noise and interference. This is rarely a problem for broadcasts on the FM band, because they use frequency modulation.

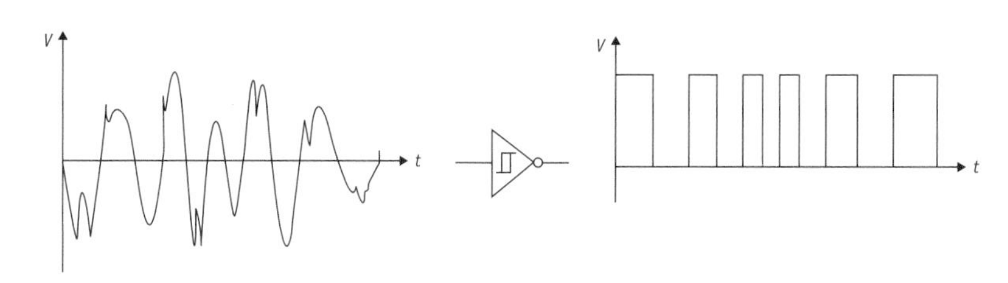

Fig 6.51 Noise can be completely removed from f.m. carriers

Questions

6.1 Amplitude modulation

1 An amplifier whose gain is variable is used to make an amplitude modulated carrier.

(a) Draw a block diagram for the system. Use the following blocks:

bias network **oscillator** **variable gain amplifier**

Label the a.f. input and the a.m. output.

(b) The oscillator is a relaxation oscillator with a frequency of 42 kHz. Draw its circuit diagram. Show all component values and justify them.

(c) Sketch voltage–time graphs to show how the a.f. signal and the h.f. carrier combine to make the a.m. carrier.

(d) Explain how the three blocks interact to make the a.m. carrier.

2 An a.f. signal of frequency 10 kHz is used to amplitude modulate an h.f. carrier of frequency 50 kHz.

(a) Show that the periods of the h.f. carrier and a.f. signal are 20 μs and 100 μs respectively.

(b) Sketch a voltage–time graph for 100 μs of the modulated carrier.

(c) Sketch the amplitude–frequency graph for the modulated carrier. Label the carrier and sidebands.

(d) The amplitude of the a.f. signal is halved and its frequency is doubled. What effect does this have on the amplitude–frequency graph of the modulated carrier?

3 Radio broadcast channels on the MW band use amplitude modulation to encode speech and music in the frequency range 50 Hz to 5 kHz.

(a) Explain the meaning of the term **amplitude modulation.**

(b) State the bandwidth required to broadcast each channel. Justify your answer with the help of an amplitude–frequency graph for one channel.

(c) Draw the circuit diagram of an a.m demodulator. Use only these components.

capacitor **diode** **resistor**

(d) Explain how the demodulator operates. Include sketches of voltage–time graphs at the input and output.

(e) State component values for the demodulator and justify them.

6.2 Pulse width modulation

Assume that the outputs of the op-amps in these questions saturate at ±13 V.

1 A pulse width modulator is made from a triangle waveform generator and a comparator.

(a) Draw a block diagram to show how this can be done. Label the input and output.

(b) With sketches of voltage–time graphs for the input and output of each block, explain how the system operates.

(c) The modulator is used to carry information about signals in the range 0 Hz to 5 kHz. Explain why the frequency of the triangle waveform must be at least 10 kHz.

(d) Draw the circuit diagram of a treble-cut filter with a break frequency of 5 kHz and a low frequency gain of 2. Show all component values and justify them with calculations.

(e) Explain how the treble-cut filter acts as a demodulator for the system of part (a).

2 This question is about the triangle waveform generator of Fig 6.52.

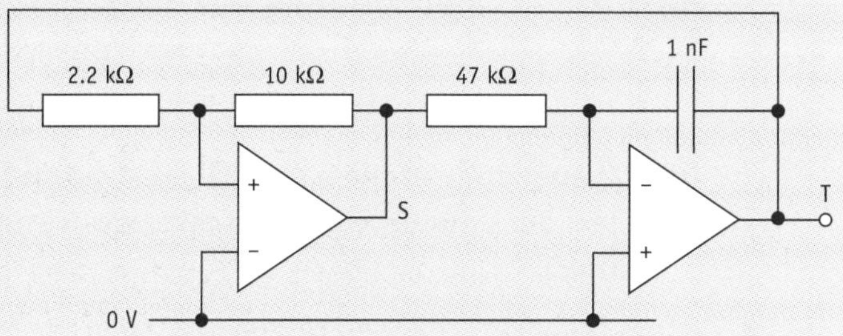

Fig 6.52 Question 2

(a) Draw a block diagram for the system, labelling S and T. Use these blocks.

Schmitt trigger ramp generator

(b) On the same set of axes, sketch graphs of two cycles of the signals at S and T.

(c) Do calculations to show that the trip points of the Schmitt trigger are at about +3 V and −3 V.

(d) Do calculations to show that the output of the ramp generator rises and falls at a rate of about 3×10^5 V s^{-1}.

(e) Calculate the frequency of the triangle waveform generator.

3 A pulse width modulator codes a signal as the mark–space ratio of a square wave whose frequency is 40 kHz.

(a) Show that the period of the square wave is 25 μs.

(b) Explain what is meant by the term **mark–space ratio.**

(c) The output of the modulator is an op-amp comparator. Sketch voltage–time graphs for two cycles of the output when the mark–space ratio is (i) 1:1 (ii) 4:1 (iii) 1:4

(d) Calculate average values for the comparator output for these mark–space ratios: 1:1, 1:4 and 4:1.

6.3 Frequency modulation

1 The variable frequency oscillator whose transfer characteristic is shown in Fig 6.53 is used to produce a frequency modulated carrier.

Fig 6.53 Question 1

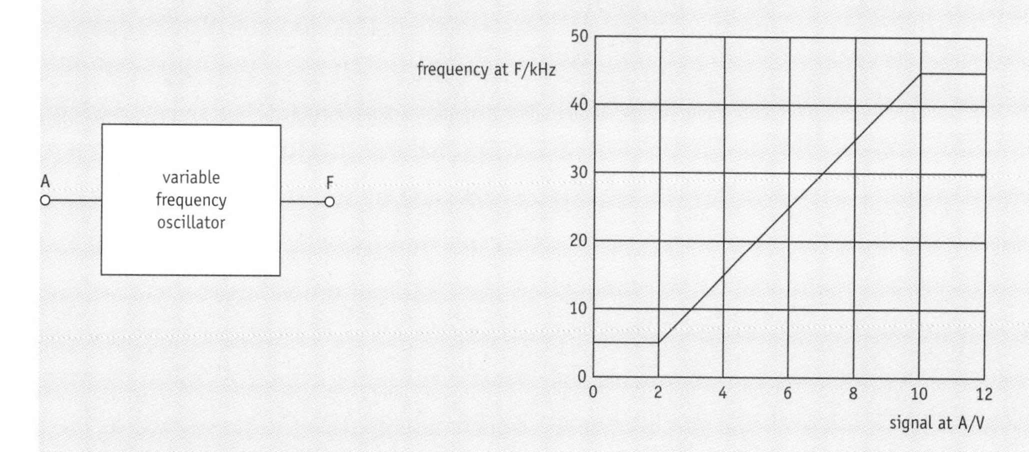

(a) Explain why A should be held +6 V by a bias network.

(b) The signal to be coded by the carrier has a frequency range of 25 Hz to 2.5 kHz. Draw a circuit diagram for the bias network. Show all component values and justify them with calculations.

(c) The signal to be coded oscillates from −1.0 V to +1.0 V. State the corresponding maximum and minimum frequencies at the output of the modulator.

(d) On the same set of axes, sketch voltage–time graphs of the signals at the input and output of the frequency modulator. Assume that the output is a sine wave with a constant amplitude of 3 V and that the input has a frequency of 2.5 kHz.

2 Radio broadcasts in the FM band use frequency modulation to transmit music in the range 15 Hz to 15 kHz.

(a) Explain what is meant by the term **frequency modulation.**

(b) Calculate the bandwidth required for each broadcast in the FM band.

(c) Draw a block diagram for a frequency demodulator. Use only these blocks.

monostable **Schmitt trigger** **filter**

(d) With a description of the behaviour of each block, explain how the demodulator operates.

3 A monostable is required for a frequency demodulator.

(a) Draw the circuit diagram of a monostable which produces a 5 V pulse each time it is triggered by a falling edge.

(b) State component values which give a pulse duration of 40 μs.

(c) Explain why the frequency of the modulated carrier must be less than 25 kHz.

(d) The signal encoded in the modulated carrier is in the range of 20 Hz to 2.0 kHz. Draw a circuit diagram for the filter in the demodulator. Show all component values and justify them with calculations.

6.4 Signal transmission

1 Modulated carriers can be transmitted from one place to another.

(a) Describe **three** different ways of transmitting a modulated carrier.

(b) During transmission, the modulated carrier picks up noise and interference. Explain what is meant by the terms **noise** and **interference**.

(c) What else happens to a modulated carrier during transmission? Explain what effect it might have on the information being transferred.

2 Modulated carriers are transmitted from one place to another along wires.

(a) Explain the advantages of using a twisted-pair cable.

The signal-to-noise ratio of the modulated carrier changes as it passes along the wires.

(b) Explain what is meant by the term **signal-to-noise ratio.**

(c) Explain the change of signal-to-noise ratio as the carrier passes along the wires.

(d) Describe what can be done to reduce the effect of this change.

3 The internet uses optical fibres to transmit information around the world.

(a) Explain how information can be sent along an optical fibre.

(b) What are the advantages of using optical fibre instead of radio waves or wires for long distance communication?

4 Frequency modulated radio waves always pick up interference on their way from transmitter to receiver.

(a) Explain how the use of Schmitt trigger usually allows the interference to be removed from the modulated carrier.

(b) Not all radio broadcasts are immune from interference. Suggest a reason why.

(c) Explain why interference is not a problem when information passes through optical fibres.

Learning summary

By the end of this chapter you should be able to:

- understand how to use a variable gain amplifier for amplitude modulation
- understand how to use a variable frequency amplifier for frequency modulation
- understand how to use a triangle waveform generator for pulse width modulation
- calculate the trip points for a Schmitt trigger
- calculate the rate at which the output of a ramp generator changes
- design demodulators for a.m., f.m. and p.w.m carriers
- calculate the bandwidths required to transmit a.m., f.m. and p.w.m. carriers
- describe the relative merits of radio waves, optical fibre and wires for transmitting modulated carriers.

CHAPTER 7

Frequency division multiplexing

7.1 Channels

Think about listening to your favourite station on the radio. Track after track is played, each one better than the last . . . until the presenter decides to interview the latest band. In exasperation, you reach for the tuning button and flick through the channels until you find another one playing your type of music. You can do this because each station broadcasting in a particular area is allocated a different range of radio frequencies (or **channel**) for its own use. So each time you tune your radio, you are selecting a different range of frequencies. For example, my own favourite station has exclusive use of all the radio frequencies between 91.25 MHz and 91.35 MHz. When it gets boring I retune to the channel centred on the frequency 100.90 MHz.

Shared link

All radio stations have to share the same transmission medium, our atmosphere. Their aerials pour out radio waves which have to share the same space as the radio waves emitted by other stations in the same locality. So the aerial of a receiver will have lots of different alternating currents as it absorbs these various radio waves. The only way of preventing confusion is to insist that each station broadcasts a unique set of frequencies, and let the receiver filter out all but one set. This technique of allowing many different users to share the same transmission link is called **frequency division multiplexing** (or **f.d.m.**), and is widely used for broadcast TV and radio. Radio frequencies extend from 30 kHz to 300 GHz, and are split into a number of different **bands**. Waves within a particular band have their own characteristics, determining their use, as shown in Fig 7.1. For example, UHF signals are line-of-sight only, and so don't go over the horizon, allowing them to be reused in different areas.

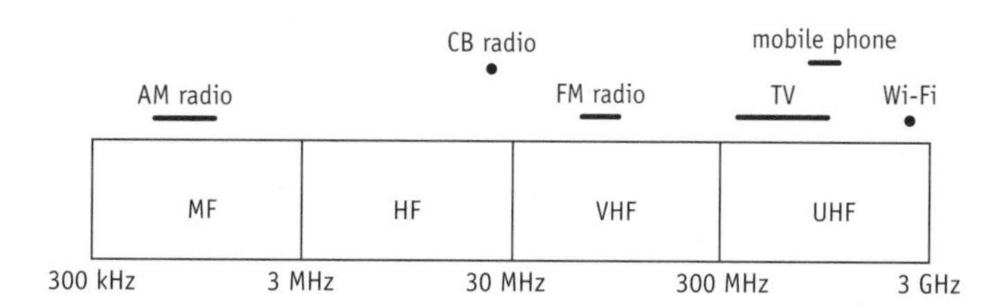

Fig 7.1 Four of the radio frequency bands

Channel number

The number of different channels that can be made available depends on the bandwidth required for each broadcast. Consider the FM band.

- It extends from 87.5 MHz to 108.0 MHz.
- Each station uses frequency modulation to encode sound in the frequency range 30 Hz to 15 kHz.
- Two stereo signals (L and R) are separately encoded as L + R and L − R. Mono receivers only decode the L+R signal.

The bandwidth required for each station is easily estimated. Start off by considering the mono signal L + R.

$\Delta f_{fm} = ?$ $\Delta f_{fm} = 5f_s = 5 \times 15\times10^3 = 7.5\times10^4$ Hz or 75 kHz

$f_s = 15$ kHz

The bandwidth required is double this to accommodate stereo, giving 150 kHz altogether. Each channel in the FM band is given a spread of 200 kHz, so that there is a clear region of 50 kHz between adjacent stations, as shown in Fig 7.2. The FM spectrum covers a range of 108.0 − 87.5 = 20.5 MHz. Each channel is allocated 0.2 MHz. So the maximum number of channels is 20.5 / 0.2 = 102.

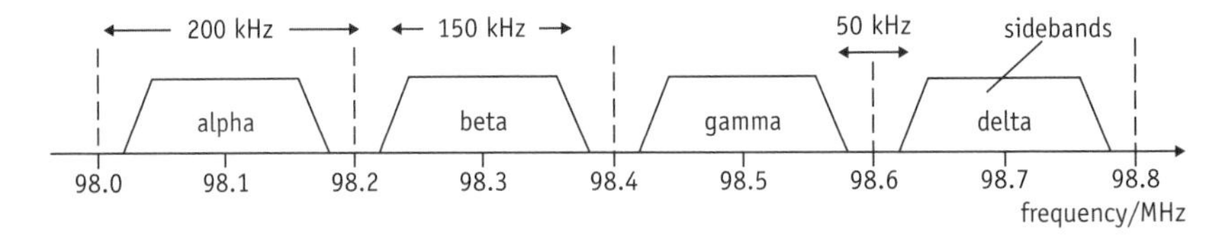

Fig 7.2 Four channels in the FM radio spectrum

Bandpass filter

The block diagram of Fig 7.3 shows the process of f.d.m. for four adjacent radio stations in the FM band. Each station (alpha, beta, gamma and delta) has its sound frequency modulated onto a different carrier. All four modulated carriers pass along the same transmission link (the atmosphere) and arrive at four different receivers. Each receiver has a different **bandpass filter**, allowing just one channel through to the demodulator. It should be obvious that, for this multiplexing scheme to work, each filter must have the following properties.

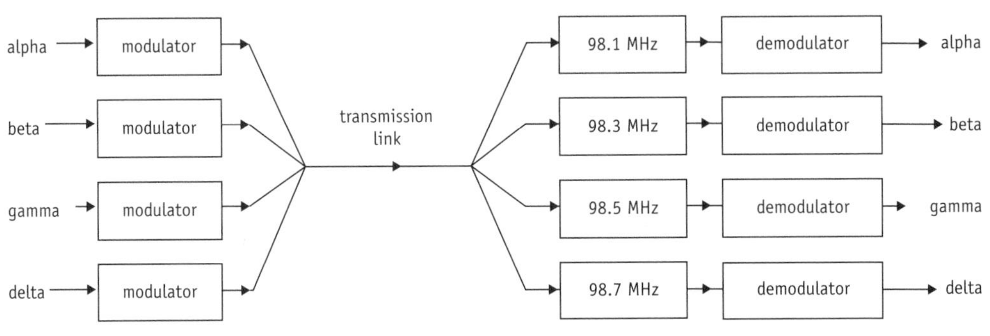

Fig 7.3 Frequency division multiplexing requires bandpass filters

- A gain of 1 for all frequencies in the channel.
- A gain of 0 for all frequencies outside the channel.

This ideal transfer characteristic is illustrated in Fig 7.4.

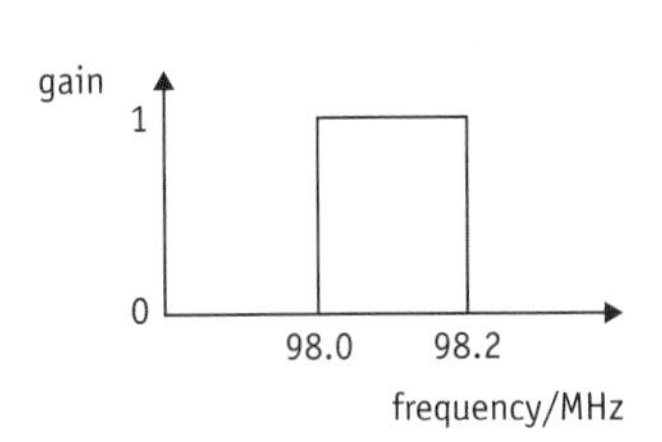

Fig 7.4 Ideal transfer characteristic for the 98.1 MHz filter

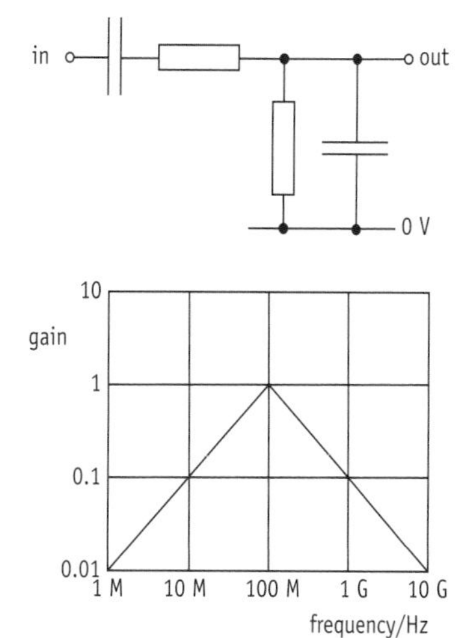

Fig 7.5 Transfer characteristic for a notch filter, using a two-line approximation

LC filter

The **notch filter** of Fig 7.5 is clearly not going to be suitable as the bandpass filter for f.d.m. Although the gain peaks at the break frequency of the RC components, it drops far too slowly on either side. Fig 7.6 shows a better circuit.

The new component in Fig 7.6 is an **inductor**, connected in parallel with a capacitor. Its construction is very simple, just a coil of wire wound around a slug of iron-laden ceramic. The transfer characteristic of the network is a big improvement on that of the notch filter but is still not ideal.

- The gain varies with frequency within the channel, reducing the amplitude of the high frequency sidebands.
- Outside the channel, the gain is not zero, so some signals from adjacent channels will get through to the demodulator.

However, as you will find out later on, stacking three of these **LC filters** together makes a device which only lets through a narrow range of frequencies.

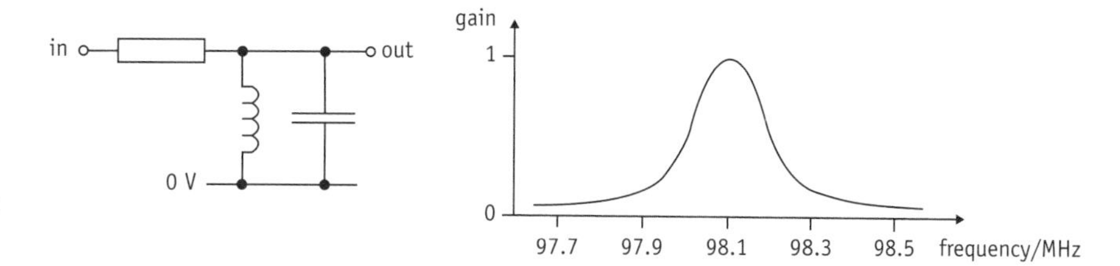

Fig 7.6 Transfer characteristic for an LC filter

Resonant frequency

An LC filter has a maximum gain at its **resonant frequency** f_0. The value of f_0 (in Hz) can be calculated from the capacitance C (in F) and the inductance L (in H) with this formula.

$$f_0 = \frac{1}{2\pi\sqrt{LC}}$$

So what is the resonant frequency of an LC filter with a capacitance of 100 pF and an inductance of 220 μH (**microhenries**)?

$f_0 = ?$

$L = 220\ \mu\text{H}$

$C = 100\ \text{pF}$

$$f_0 = \frac{1}{2\pi\sqrt{LC}} = \frac{1}{2\pi\sqrt{220 \times 10^{-6} \times 100 \times 10^{-12}}}$$

$$f_0 = 1.1 \times 10^6\ \text{Hz or } 1.1\ \text{MHz}$$

This is in the MF band, suitable for use with AM radio receivers. So what value of inductance would you need to get a resonant frequency of 98 MHz in the FM band? You will need a small capacitor, so let's try 2.2 pF.

$L = ?$

$f_0 = 98\ \text{MHz}$

$C = 2.2\ \text{pF}$

$$f_0 = \frac{1}{2\pi\sqrt{LC}} = 1.6 \times 10^{-9} \qquad \sqrt{LC} = \frac{1}{2\pi f_0} = 1.6 \times 10^{-9}$$

$$L = \frac{(1.6 \times 10^{-9})^2}{2.2 \times 10^{-12}} = 1.2 \times 10^{-6}\ \text{H or } 1.2\ \mu\text{H}$$

This inductor corresponds to about 30 turns of wire coiled loosely around a pencil!.

Reactance

In order to understand the operation of an LC filter, you need to know about **reactance**. This is a component's voltage-to-current ratio when it carries an alternating current, a measure of its resistance to a.c. signals. More exactly, you can calculate the reactance X of a component (in Ω) from its peak voltage V_0 (in V) and peak current I_0 (in A) with this formula.

$$X = \frac{V_0}{I_0}$$

It is no accident that this looks similar to the formula for resistance. Indeed, for a resistor, the reactance X_R is the same as its resistance R, as shown in the table.

component	reactance
resistor R	$X_R = R$
capacitor C	$X_C = \frac{1}{2\pi fC}$
inductor L	$X_L = 2\pi fL$

Look at the formulae in the table. The reactance of each component responds differently to an increase of frequency f.

- A resistor's reactance does not change.
- A capacitor's reactance decreases with increasing frequency.
- An inductor's reactance increases with increasing frequency.

It is this difference in behaviour which gives an LC filter its bandpass characteristic. To understand this, consider a 220 μH inductor in parallel with a 100 pF capacitor, as shown in Fig 7.7.

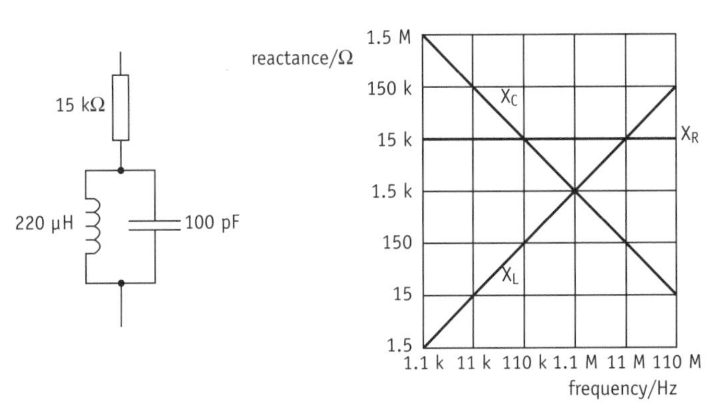

Fig 7.7 Variation of reactance with frequency for the components of an LC filter

Each component has its own line on the reactance–frequency plot. To draw it, you only need to calculate the reactance for one frequency. For example, what is the reactance of the capacitor at a frequency of 11 kHz?

$X_C = ?$

$f = 11$ kHz

$C = 100$ pF

$$X_C = \frac{1}{2\pi fC} = \frac{1}{2\pi \times 11 \times 10^3 \times 100 \times 10^{-12}}$$

$$X_C = 1.5 \times 10^5\ \Omega \text{ or } 150\ \text{k}\Omega$$

As you can see from the graph, the line for X_C goes though this point, dropping by a factor of ten for every tenfold increase of frequency.

Now consider the reactance of the inductor at the same frequency of 11 kHz.

$X_L = ?$ $\qquad X_L = 2\pi fL = 2\pi \times 11\times10^3 \times 220\times10^{-6} = 15\ \Omega$

$f = 11$ kHz

$L = 220\ \mu$H

So the line for X_L must go through this point and rise by a factor of ten for every tenfold increase of frequency.

Impedance

The **impedance** of the parallel LC circuit is a measure of its total resistance to a.c. signals. The graph of Fig 7.8 shows how the impedance Z is determined by the frequency. Since the inductor and capacitor are in parallel, the greater current passes through whichever has the lower reactance.

- Below 1.1 MHz, almost all of the current is in the inductor, so Z is the same as X_L.
- Above 1.1 MHz, almost all of the current is in the capacitor, so Z is the same as X_C.

The big surprise comes at the resonant frequency, where both components have the same reactance. You would expect the current to split equally between them, to give an impedance equal to half of 1.5 kΩ. Instead, the impedance shoots up to 100 times this value, reaching 150 kΩ. This means that, at the resonant frequency, the LC circuit has a much larger impedance than the resistor in series with it, giving the filter a gain of 1. At all other frequencies, Z is much smaller than R, giving the filter a gain of almost 0.

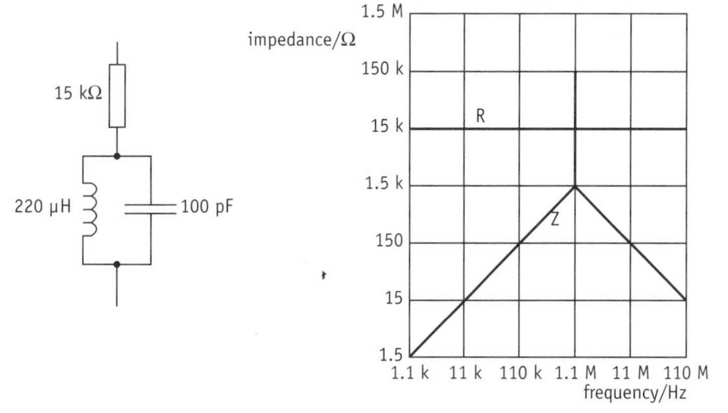

Fig 7.8 The impedance Z rises above the resistance R at the resonant frequency

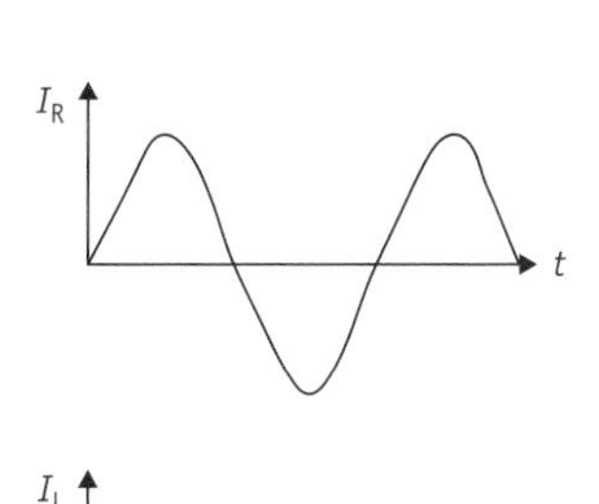

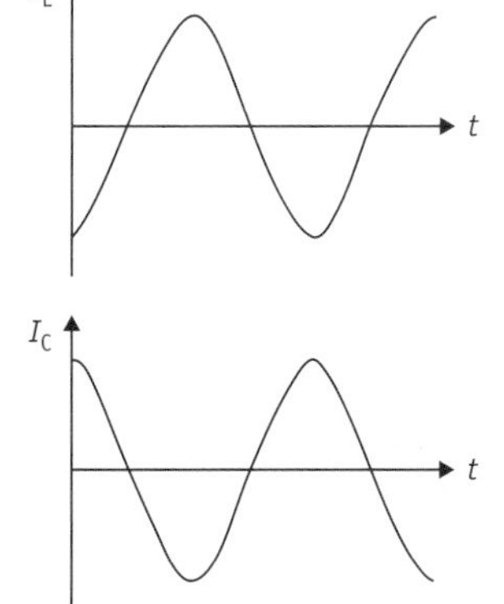

Fig 7.9 The currents have different phases

Phase

So why does the impedance of the LC circuit change so dramatically at the resonant frequency? It all has to do with the **phase** of the current in each component. Take a look at the three graphs of Fig 7.9.

Notice how the peak current occurs at different times for each component, even though the peak voltage happens at the same time. In particular, the current in the capacitor and inductor have opposite signs, allowing them to partially cancel. This means that the current in the resistor is always less than the current in either the inductor or the capacitor. At the resonant frequency, when $X_C = X_L$, the currents cancel out almost completely. This results in almost no current in the resistor, making the parallel LC circuit look like a very large resistor.

Bandwidth

The impedance of a parallel LC circuit at its resonant frequency depends on the resistance of the inductor. (Since an inductor is a coil of wire, it will always have some resistance, however small.) Increasing the resistance of the inductor affects the filter in two ways.

- It reduces the impedance at the resonant frequency.
- It increases the bandwidth of the filter.

This means that the shape of the gain–frequency graph can be changed by altering the resistance *r* of the LC circuit, as shown in Fig 7.10. It is clearly important to get the bandwidth matched to the width of the channel. Too small a bandwidth means a loss of high frequency sidebands, distorting the demodulated signal. Too large a bandwidth allows sidebands of neighbouring channels through to the demodulator, introducing interference. The bandwidth needs to be just right.

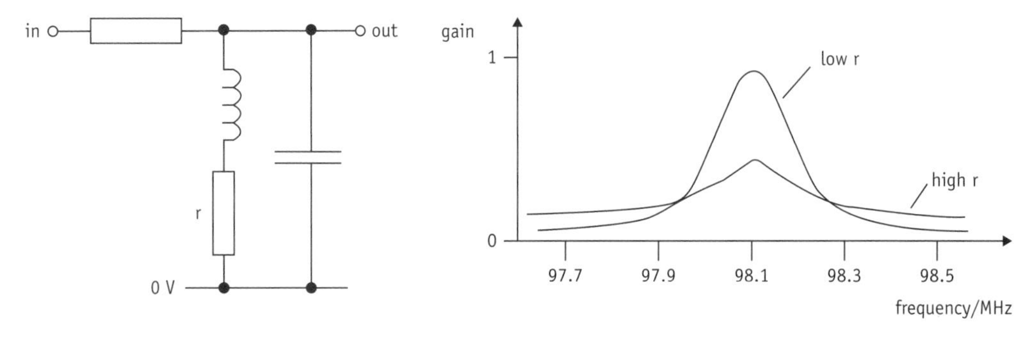

Fig 7.10 The resistance of the inductor determines the bandwidth and gain of the filter

Stacked filters

A single parallel LC filter often isn't much use on its own. Its gain–frequency curve doesn't have the flat top and sharply dropping edges required for a useful band-pass filter. Only when three of them are put in series, as shown in Fig 7.11, do you get a useful filter which is capable of selecting just one channel out of many.

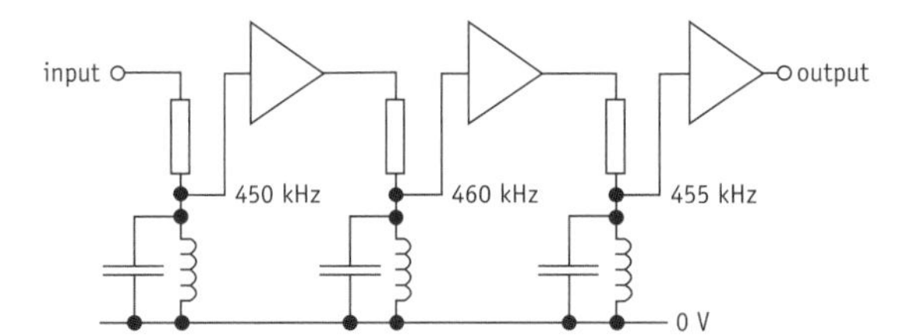

Fig 7.11 Each LC filter in this stack of three has a slightly different resonant frequency

Notice two features of the **filter stack**.

- Each LC filter has a slightly different resonant frequency.
- There is an **amplifier** at the output of each filter.

The correct choice of resonant frequency and bandwidth for each filter gives the overall transfer characteristic of Fig 7.12. Signals between 450 kHz and 460 kHz get through with hardly any change, but are severely attenuated outside this range. Getting the correct bandwidth for each filter, as well as its resonant frequency, is crucial for the success of the whole circuit. This is why the amplifiers have been included. They provide each filter with a high impedance load, so that its characteristic is not affected by any current drawn from it.

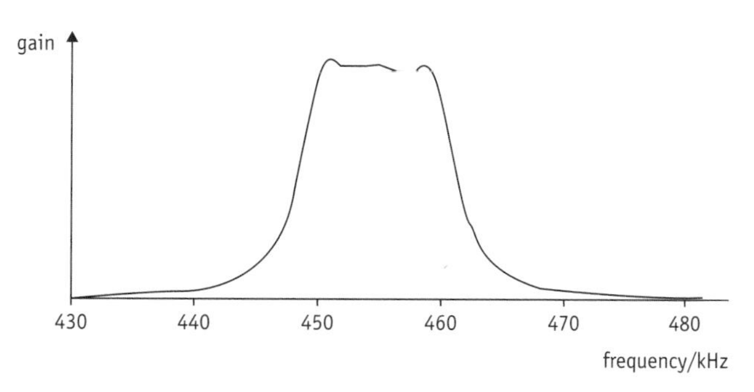

Fig 7.12 Gain–frequency curve for the stacked filter of Fig 7.11

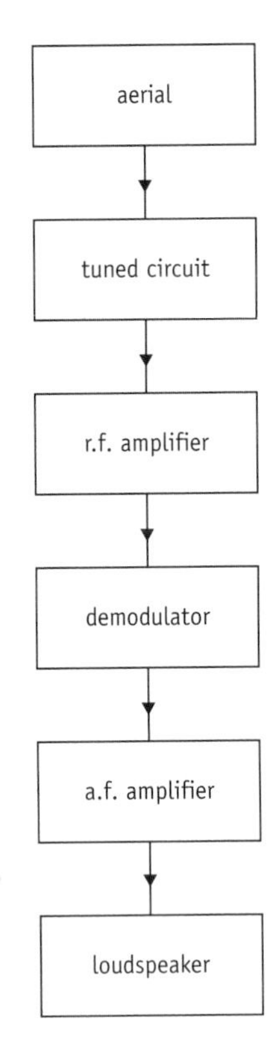

Fig 7.13 Block diagram for a simple radio receiver

7.2 Radio receivers

Even the simplest radio receiver is actually quite complex. Think of what it has to do.

- Absorb radio waves to set up modulated alternating currents in the aerial.
- Reject all frequencies that are outside the selected channel.
- Demodulate the **radio frequency** (or **r.f.**) signal to recreate the a.f. signal.
- Amplify the a.f. signal and feed it to a loudspeaker to recreate speech or music.

The block diagram of such a system is shown in Fig 7.13. This section is going to explain the function of each stage of a simple radio receiver, and show how its shortcomings can be overcome with a more sophisticated design.

Input sub-system

The input sub-system of every radio receiver, including the one in your mobile phone, consists of just three passive components:

- An **aerial** which absorbs radio waves and sets up alternating currents.
- A **tuned circuit** made from a capacitor and inductor in parallel.

The arrangement is shown in Fig 7.14.

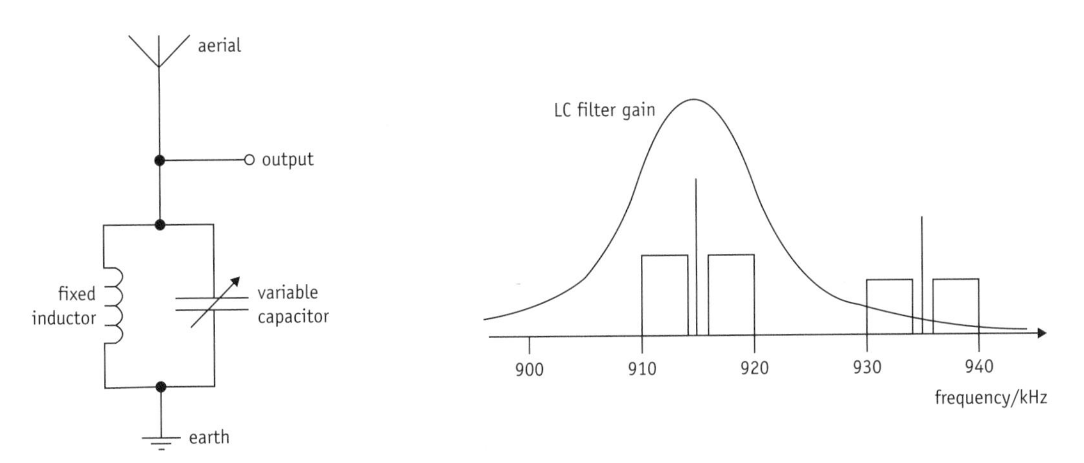

Fig 7.14 The tuned circuit needs to be centred on one channel

Notice that the bottom end of the circuit has been **earthed**. Although this is not strictly necessary (and not even possible in a mobile phone) it usually increases the current in the aerial. As the radio waves are absorbed by the aerial, they set up high frequency alternating currents in the parallel LC circuit as charge flows between the aerial and the ground. For most frequencies, the LC circuit has a very low impedance, so the current generates only a small voltage across it. The LC circuit only has a high impedance for alternating currents which are close to its resonant frequency. Only these currents will therefore set up a relatively large alternating voltage across the LC circuit, leading to a small a.m. carrier at the output – usually with an amplitude of less than a few millivolts.

Selectivity

Fig 7.14 shows the frequency spectrum of two adjacent channels. Each is shown as a central carrier (the spike) with a set of sidebands (the rectangles) covering the range of a.f. signals on either side. Although the peak of the LC filter's gain–frequency curve has been centred on the channel at 915 kHz, the gain is still high enough over adjacent channels to let them through to the demodulator. This means that information from more than one station can easily get through to the loudspeaker, resulting in a jumble of sound. In other words, the receiver lacks **selectivity**. It will not always be able to select signals from just one channel. Of course, you can always reduce the bandwidth of the LC filter by reducing the resistance of the inductor, but you then run the risk of attenuating the high frequency sidebands. As you will find out later on, the use of mixing and a stacked filter in a superhet receiver gets around this problem.

Tuning

One of the components in the tuned circuit needs to be variable, so that the resonant frequency of the parallel LC circuit can be matched to one of the channels being broadcast. It is usually the capacitor, indicated by an arrow through its symbol in Fig 7.14. For example, a popular variable capacitor can have any value between 5 pF and 120 pF, simply by rotating a shaft. If it is placed in parallel with a 1 mH inductor, what range of resonant frequencies can the tuned circuit have? Consider the value of f_0 when the capacitor is set to 5 pF.

$f_0 = ?$

$L = 1$ mH

$C = 5$ pF

$$f_0 = \frac{1}{2\pi\sqrt{LC}} = \frac{1}{2\pi\sqrt{1 \times 10^{-3} \times 5 \times 10^{-12}}}$$

$$f_0 = 2.3 \times 10^6 \text{ Hz or } 2.3 \text{ MHz}$$

The lower end of the range happens for the largest value of the capacitor.

$f_0 = ?$

$L = 1$ mH

$C = 120$ pF

$$f_0 = \frac{1}{2\pi\sqrt{LC}} = \frac{1}{2\pi\sqrt{1 \times 10^{-3} \times 120 \times 10^{-12}}}$$

$$f_0 = 4.6 \times 10^5 \text{ Hz or } 460 \text{ kHz}$$

This comfortably covers the entire range of AM broadcasts in the MF band (530 kHz to 1.61 MHz).

Loading

A tuned circuit works best when it is left alone. As soon as you connect the output to some other system, two things happen.

- The bandwidth of the LC filter increases, making the receiver less selective.
- The amplitude of the signal at the output decreases. This makes it more difficult for the receiver to pick up weak radio signals, making it less **sensitive**.

As you will find out on the next page, a MOSFET voltage follower can deal with this problem. Since there is no current in the gate, it cannot affect the performance of the LC filter.

Radio frequency

The signal at the base of the aerial is small, rarely with an amplitude of more than 10 millivolts. This needs to be boosted in two stages before it enters the demodulator.

- A high input impedance voltage follower extracts the signal from the tuned circuit.
- An inverting amplifier increases the amplitude of that signal.

A circuit which does this is shown in Fig 7.15.

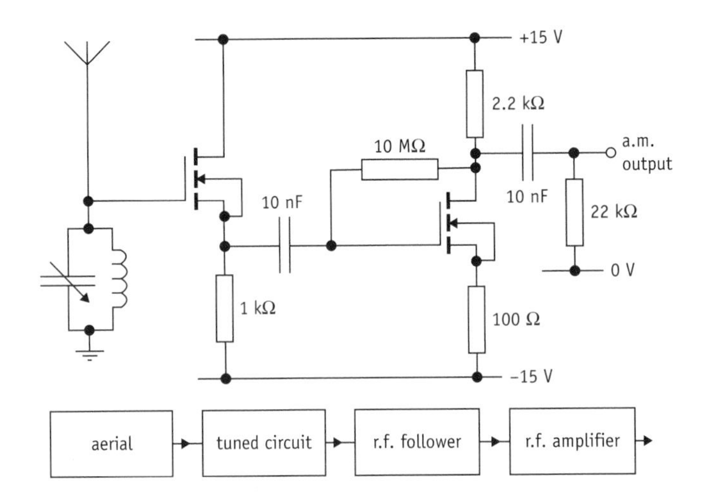

Fig 7.15 Circuit diagram for the r.f. blocks

Notice how the tuned circuit provides the bias for the voltage follower, keeping the average voltage of the gate at 0 V. As the gate moves up and down in voltage, so does the source. The r.f. amplifier has a gain of −22, increasing the amplitude of the r.f. signal and changing its sign. The latter makes it less likely that feedback from the a.m. output to the aerial will result in oscillation.

Demodulator

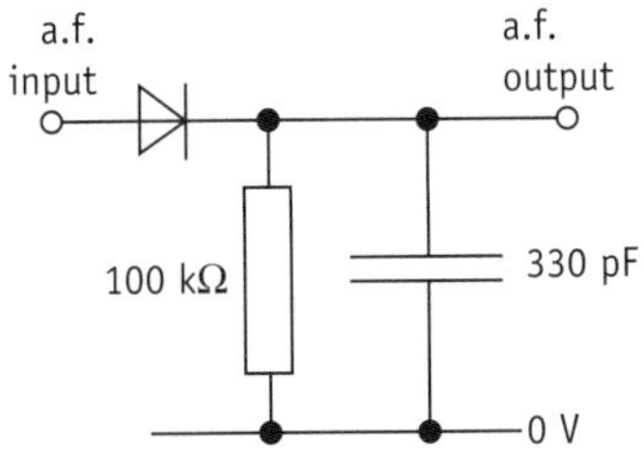

a.f. input
a.f. output
100 kΩ
330 pF
0 V

Fig 7.16 Demodulator circuit

Since radio broadcasts in the MF band use amplitude modulation, their a.m. carriers can be demodulated with a diode, resistor and capacitor, as shown in Fig 7.16. The resistor needs to be much larger than the output impedance of the previous stage. As you can see from Fig 7.15, this is likely to be between 2.2 kΩ and 22 kΩ, so 100 kΩ should be all right. The value of the capacitor can be calculated by setting the break frequency to the highest frequency expected in the a.f. signal. For AM broadcasts in the MF band, this is 5 kHz.

$C = ?$

$R = 100\ \text{k}\Omega$

$f_0 = 5\ \text{kHz}$

$$f_0 = \frac{1}{2\pi RC}$$

$$C = \frac{1}{2\pi R f_0} = \frac{1}{2\pi \times 100 \times 10^3 \times 5 \times 10^3}$$

$$C = 3.2 \times 10^{-10}\ \text{F or } 320\ \text{pF}$$

The nearest value of 330 pF has been used in Fig 7.16. This will let the a.f. signal pass through unaffected, yet attenuate the r.f. components at 1 MHz by at least $5\times10^3/1\times10^6 = 0.005$.

Audio frequency

The a.f. signal extracted from the a.m. carrier by the demodulator is going to be quite small. It therefore needs some processing before it can be used to create speech and music via a loudspeaker. Suitable circuitry is shown in Fig 7.17.

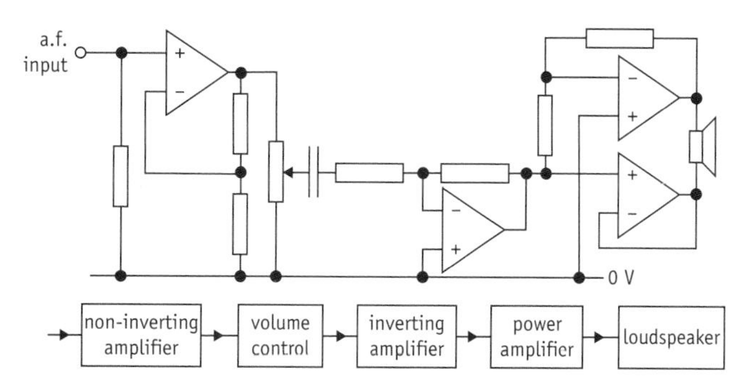

Fig 7.17 Circuit diagram for the a.f. blocks

Notice the order of the subsystems.

- A high input impedance non-inverting amplifier to extract and boost signals from the demodulator without affecting its operation.
- A volume control with a.c. coupling at its output to eliminate the d.c. component from the demodulator.
- A high gain inverting amplifier. The negative gain makes it less likely that feed-back from output to input will set the system into oscillation.
- A push-pull power amplifier to increase the current for the loudspeaker.

Of course, none of this is much good if the a.f. signal from the demodulator contains information from more than one channel. You only want to listen to one radio broadcast at any one time!

Superhet

The simple radio receiver discussed in the last four pages suffers from one major drawback; it lacks selectivity. This is because it uses only one tuned circuit, so that it can never satisfactorily separate one channel from its neighbour. This is avoided in the **superhet** receiver whose block diagram is shown in Fig 7.18.

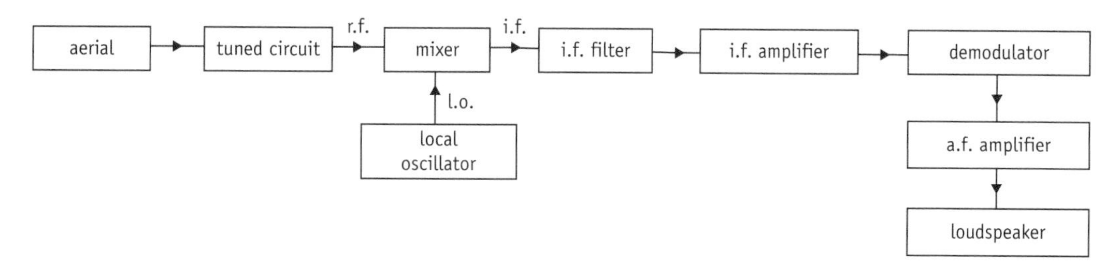

Fig 7.18 Block diagram for a superhet radio receiver

Notice the block labelled **i.f. filter**. It is a stacked filter, centred on the **intermediate frequency** (or **i.f.**) of 455 kHz, with a bandwidth of 10 kHz. This is just wide enough to pass one a.m. carrier and its sidebands if the a.f. signal is restricted to less than 5 kHz. It is this filter, made from three tuned circuits in a row, which allows the superhet receiver to select signals from one channel and reject them from all the others.

Mixing

The **mixer** of a superhet receiver uses the signal from the **local oscillator** (or **l.o.**) to amplitude modulate the **radio frequency** (or **r.f.**) signal from the tuned circuit. You will recall that amplitude modulation adds a pair of sidebands to the frequency spectrum of the carrier, according to this rule.

$$f_{\text{sideband}} = f_{\text{carrier}} \pm f_{\text{modulation}}$$

So let's suppose that the tuned circuit has been centred on the channel at 845 kHz. The oscillator which generates the l.o. signal is **ganged** to the tuned circuit, so that its frequency is always 455 kHz higher. So the l.o. signal is at 845 + 455 = 1300 kHz. The frequencies of the upper and lower sidebands created by the mixer are calculated as follows.

$$f_{\text{upper}} = 845 + 1300 = 2145 \text{ kHz}$$

$$f_{\text{lower}} = 845 - 1300 = 455 \text{ kHz}$$

Notice that since a negative frequency is impossible (it requires time to run backwards!), the sign of the calculation is ignored. Of course, it is not only the carrier at 845 kHz that acquires a pair of sidebands. Each sideband of the a.m. carrier centred on 845 kHz also gets its own sidebands which are 455 kHz higher and lower than itself. So the mixer creates an extra pair of copies of the a.m. carrier at 845 kHz, centred on 2145 kHz and 455 kHz, as shown in Fig 7.19.

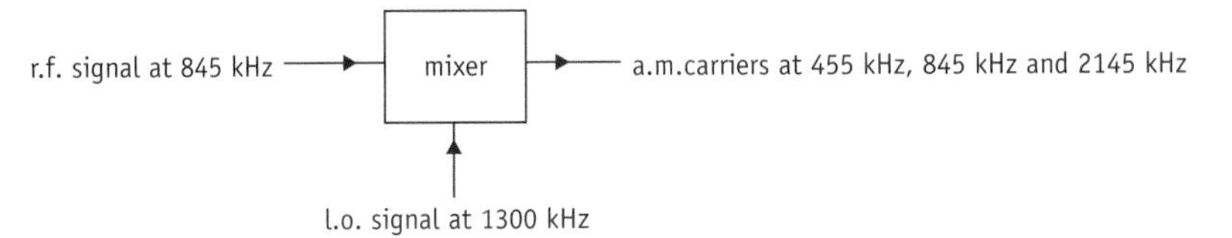

Fig 7.19 The mixer creates two extra copies of the a.m. carrier at its input

Superhet operation

The frequency spectra of Fig 7.20 on the next page illustrate how the superhet receiver manages to select just the one channel at 845 kHz.

- The r.f. signal from the tuned circuit shows three different stations around 845 kHz.
- The l.o. signal is a single spike, 455 kHz above the centre frequency of the tuned circuit.
- The mixer amplitude modulates the r.f. signal with the l.o. signal to create the i.f. signal. This contains copies of the r.f. signal centred on 455 kHz and 2145 kHz.
- The i.f. bandpass filter lets through only the signals between 450 kHz and 460 kHz. This will be the sidebands of the station at 845 kHz on either side of a carrier at 455 kHz.

So the act of tuning a superhet receiver really involves two changes. The tuned circuit has to be centred on the required station, and the local oscillator needs to be set to a frequency which is 455 kHz higher than the carrier of that station. This ensures that the output of the mixer contains a copy of the a.m. carrier which can get through the i.f. filter to the demodulator, a.f. amplifier and loudspeaker.

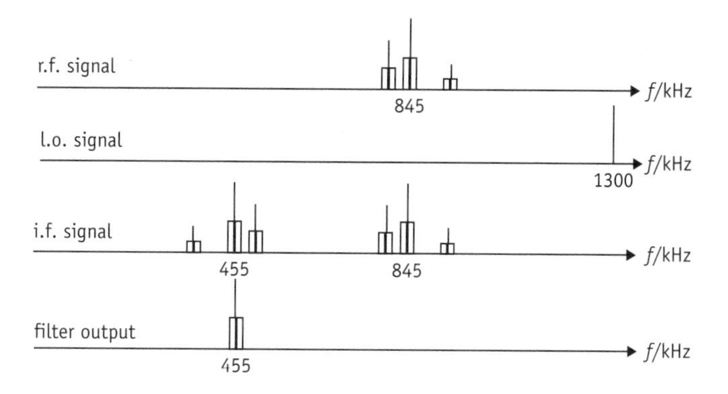

Fig 7.20 Frequency spectra for signals in a superhet receiver

Intermediate frequency

So how does the output of the local oscillator manage to stay exactly 455 kHz above the centre frequency of the tuned circuit? One solution is shown in Fig 7.21.

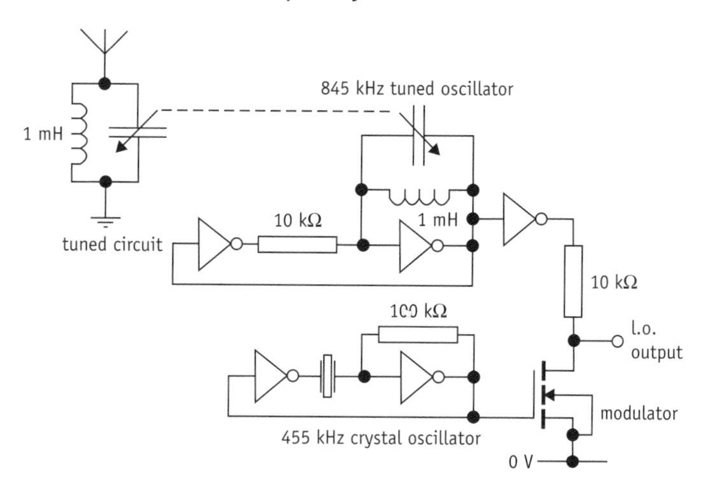

Fig 7.21 Circuit diagram for the local oscillator

The circuit has three sub-systems.

- A tuned oscillator built around a copy of the parallel LC circuit at the base of the aerial. The variable capacitor is ganged to its twin at the base of the aerial, so this subsystem oscillates at the r.f. frequency.
- A crystal oscillator which produces a square wave with a fixed frequency of 455 kHz.
- An amplitude modulator which modulates the r.f. frequency at 455 kHz.

The output of this local oscillator contains signals at three different frequencies, the r.f. frequency and a pair of sidebands which are 455 kHz above and below. A little thought should convince you that an l.o. signal which is 455 kHz below the r.f. frequency should work just as well as one which is above it. There is therefore no need to filter it out before it gets to the mixer.

Questions

7.1 Channels

1 Radio stations which broadcast on the MF band (between 530 kHz and 1.61 mHz) are each allocated a channel which is 20 kHz wide.

(a) Calculate the maximum number of stations which could broadcast on the MF band in a given area.

(b) Each station uses amplitude modulation to encode voice signals between 40 Hz and 5 kHz onto a carrier. Calculate the bandwidth required to transmit each station.

(c) Explain why the range of frequencies allocated to each channel is greater than the bandwidth required for transmission.

2 A bandpass filter is made from an inductor, a capacitor and a resistor.

(a) Draw a circuit diagram for the filter. Label the input and output.

(b) The inductor has a reactance of 10 Ω at a frequency of 10 kHz. Draw a log-log graph to show how its reactance changes over the frequency range 1 kHz to 10 MHz.

(c) The capacitor has a reactance of 100 Ω at a frequency of 10 MHz. On the log-log graph of part (b), show how its reactance changes with frequency.

(d) Use the data of (b) and (c) to calculate values for the inductor and capacitor.

(e) Use the log-log graph to describe the operation of the bandpass filter.

3 A 470 μH inductor is connected in parallel with a 220 pF capacitor.

(a) Calculate the resonant frequency of the circuit.

(b) A resistor is added to make a bandpass filter. Draw the circuit diagram for the filter.

(c) Explain why signals should be extracted from the filter by a high input impedance voltage follower.

(d) Sketch a gain–frequency graph for the bandpass filter. Describe how the graph changes when the resistance of the inductor is increased.

4 A stack of three parallel LC circuits can be used to make a bandpass filter with near-ideal performance.

(a) Draw a circuit diagram for the filter. Explain why the circuit requires amplifiers as well as parallel LC circuits.

(b) Explain how the response of a stack of filters can be better than that of just one.

7.2 Radio receivers

1 The front-end of a radio receiver consists of an aerial and tuned circuit.

(a) Draw a circuit diagram for the front-end of the receiver.

(b) By describing how the impedance of the tuned circuit varies with frequency, explain how the circuit selects one radio station and rejects the others.

(c) The inductor has a fixed value of 100 μH. Calculate the value of the variable capacitor required to select a radio station broadcasting at 940 kHz.

(d) Explain the advantages of extracting the signal from the circuit with a high input impedance amplifier.

2 **(a)** Draw a block diagram for an a.m. radio receiver. Use only these blocks.

aerial   a.f. amplifier   diode demodulator   loudspeaker
r.f. amplifier   tuned circuit

(b) Describe the function of each block.

(c) Explain why the system lacks selectivity.

(d) Explain which sub-system improves the sensitivity of the system.

3 **(a)** Draw a block diagram for a superhet radio receiver. Use only these blocks.

aerial   a.f. amplifier   demodulator   i.f. amplifier   i.f. filter
local oscillator   loudspeaker   mixer   tuned circuit

(b) Explain the relationship between the settings of the tuned circuit and local oscillator.

(c) Describe the function of the mixer.

(d) Explain how the system is able to select just one broadcast station.

(e) Draw a circuit diagram for the a.m. demodulator. Give all component values and justify them. Assume an i.f. of 455 kHz and a.f. signals of between 300 Hz and 3.4 kHz.

Learning summary

By the end of this chapter you should be able to

- understand the operation of a parallel LC circuit as a filter
- use a stack of three filters to provide a perfect response
- know how to estimate the number of channels in a frequency band
- understand the operation of a simple a.m. radio receiver
- understand the operation of a superhet radio receiver

CHAPTER 8

Time-division multiplexing

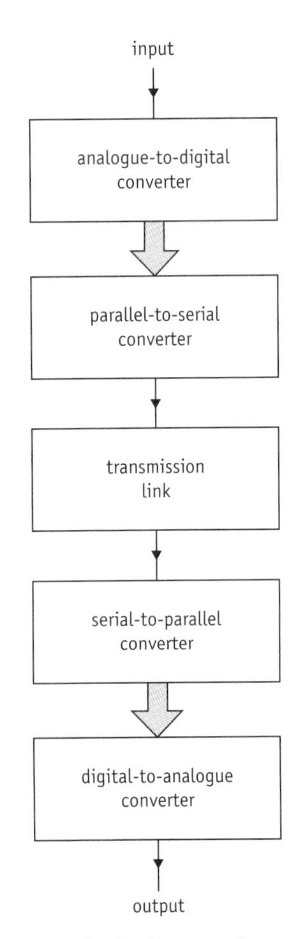

Fig 8.1 Block diagram for a digital transmission system

8.1 Digital transmission

We are a social species. By and large, we work together for our mutual profit. Communication between people, in all its aspects, can be viewed as both the glue which binds us together and the lubricant which sorts out our differences. Humans exchange signals directly in many ways. We talk, write, smile and sing – all analogue signals. However, indirect exchange of signals is increasingly quite different. We use phones, email, videos and CDs to convey our thoughts – all relying on transfer of information by digital signals. So there is a rapidly growing branch of electronics devoted to transferring analogue signals from one place to another by digital means.

Bit by bit

Fig 8.1 shows the block diagram for a basic digital transmission system, the subject of this chapter.

Each of the five blocks has its own function.

- The **analogue-to-digital converter** (or **ADC**) encodes information at the analogue input signal as a steady stream of binary words. So a voice message might be encoded as 6 800 bytes per second, emerging on an 8-bit bus.
- The **parallel-to-serial converter** (or **PSC**) takes each binary word at its input and feeds it onto the link one bit at a time.
- The **transmission link** transfers digital signals from one place to another as a series of bits one after the other. The voice signal could require as much as 68 000 bits per second.
- The **serial-to-parallel converter** (or **SPC**) takes bits one after the other from the link and stacks them together to recreate the binary words.
- The **digital-to-analogue converter** (or **DAC**) converts the binary word at its input into an equivalent analogue signal at its output.

So even a humble telephone message involves a lot of processing to make it happen. However, the benefits of digital transmission usually outweigh the costs of all this extra processing circuitry.

Benefits of binary

So why is digital transmission such a good thing? Here are some of the reasons.

- Noise and interference acquired in the transmission link can usually be completely eliminated with a Schmitt trigger at the input of the SPC.
- The system can transfer more than one type of analogue signal. Voice, video and text can all be converted into streams of bits.
- Once the information is encoded in digital form it can be manipulated in a number of useful ways. It can be compressed, stored, edited and so on.
- Many different streams of digital data can be simultaneously squeezed down a single link. This is especially useful for long-distance communication, where the cost of the link is far greater than the cost of the processors at either end.

These advantages mean the digital way is increasingly becoming the only way of getting information from one place to another.

Coding compromises

Successful transmission of an analogue signal through a digital medium requires a careful choice of parameters for the DAC. Take a look at Fig 8.2.

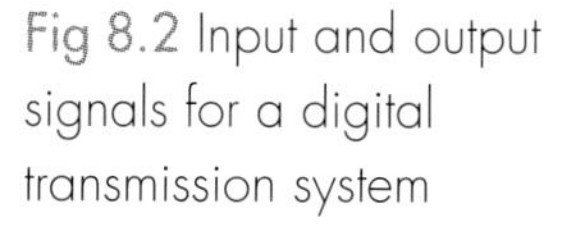
Fig 8.2 Input and output signals for a digital transmission system

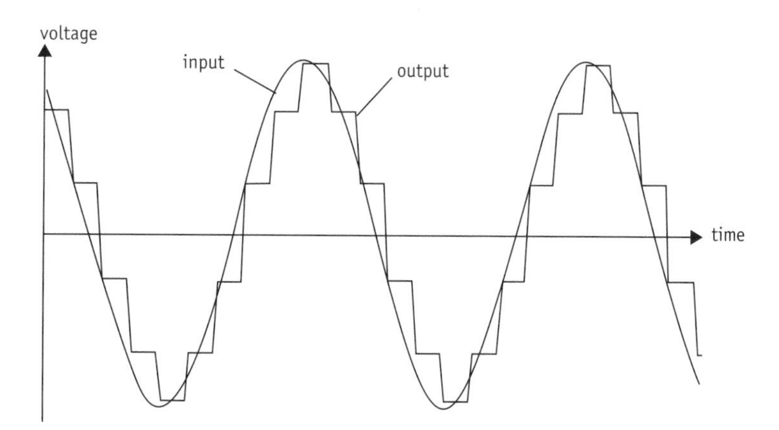

The output signal has the same amplitude and frequency as the input, but otherwise is quite different. Whereas the input moves smoothly from one voltage to another, the output goes in steps, corresponding to the eight different levels of the DAC. The input has only been **sampled** ten times in each cycle, and the output has only six different levels, so it must be **coded** as a 3-bit word ($2^3 = 8$), leading to quite a lot of distortion. Matters can obviously be improved in two ways.

- Increase the sample rate, so that the output signal is updated more frequently.
- Increase the word length, so that the interval between output levels is smaller.

Both of these require an increase in the rate at which bits must pass along the link, impacting on the number of different signals which can use that link at any one time. So digital transmission always results in some loss of information about the input analogue signal. You only send as many bits as you can get away with.

ADC operation

The timing diagram of Fig 8.3 summarises the operation of an ADC. Each time a rising edge arrives at the **sample** input, the voltage at the **input** terminal is coded as a binary word. This process takes time, so the **ready** signal goes low to signify to other systems that the binary word is being prepared. As soon as the word is safely latched onto the **output** bus, **ready** goes high again.

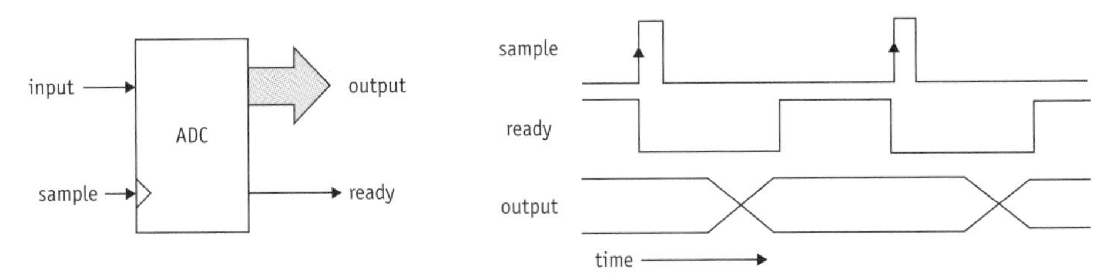

Fig 8.3 Timing diagram for an analogue-to-digital converter

Flash conversion

The fastest type of ADC is known as a **flash converter**. Its block diagram is shown in Fig 8.4, with an equivalent circuit diagram on the next page (Fig 8.5). You ought to refer to both as you work your way through the explanation below.

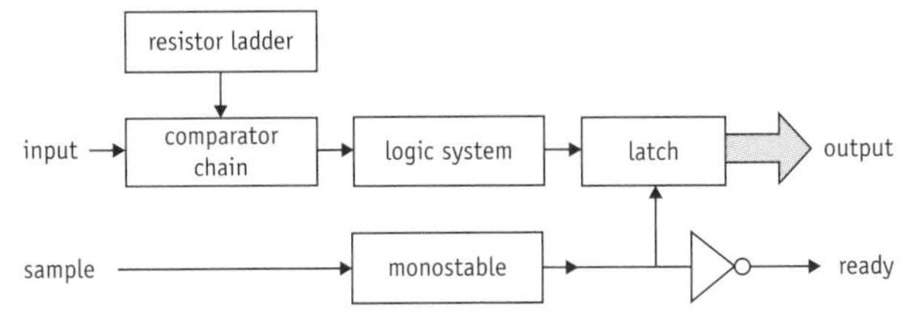

Fig 8.4 Block diagram for a flash converter

Here is the function of each block.

- The **resistor ladder** sets up a set of reference voltages.
- The **comparator chain** uses a set of op-amps to compare the voltage at **input** with each of the reference voltages. You need one op-amp for each reference voltage.
- The **logic system** takes the signals from the op-amp outputs and combines them to generate the binary word required at the output.
- The **monostable** produces a brief high pulse each time it is triggered by a rising edge at the **sample** input.
- The **latch** allows the output of the logic system through to the **output** bus while the output of the monostable is high. The output **ready** goes high to indicate that the word on the **output** bus is stable.

Look at the circuit diagram of Fig 8.5. All five resistors in the ladder have the same value, so the non-inverting inputs of each op-amp are held at the voltages shown.

input range	W	X	Y	Z
4 V – 5 V	low	low	low	low
3 V – 4 V	low	low	low	high
2 V – 3 V	low	low	high	high
1V – 2 V	low	high	high	high
0 V – 1 V	high	high	high	high

Each EOR gate, except for the top one, compares the output of each op-amp with the one above it, only going high when its inputs are different. So only one of the EOR outputs D_4 to D_1 can be 1 at any one time. The OR gates combine these outputs to generate the binary word CBA in an obvious code, as shown in the table below.

input range	D_1	D_2	D_3	D_4	C	B	A
4 V – 5 V	0	0	0	1	1	0	0
3 V – 4 V	0	0	1	0	0	1	1
2 V – 3 V	0	1	0	0	0	1	0
1 V – 2 V	1	0	0	0	0	0	1
0 V – 1 V	0	0	0	0	0	0	0

Fig 8.5 A 3-bit flash converter

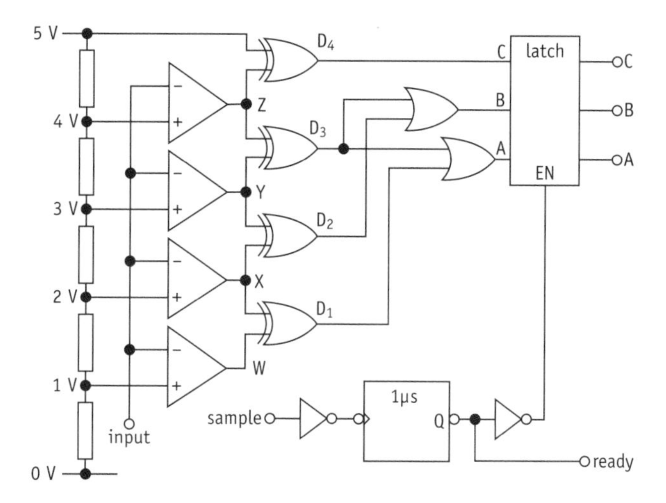

Range and resolution

The ADC of Fig 8.5 has three important parameters.

- A **conversion time** of 1 μs.
- A **range** from 0 V to 5 V.
- A **resolution** of only 1 V.

The number of **levels** is easily calculated from the last two parameters.

$$\text{levels} = \frac{\text{range}}{\text{resolution}} = \frac{5}{1} = 5$$

This requires a 3-bit word as the output. $2^2 = 4$ is not enough, so $2^3 = 8$ will have to do. The resolution can be improved by adding more op-amps, resistors and logic gates. So a 4-bit output could have $2^4 = 16$ levels, requiring another eleven op-amps. Although the design is easy to expand to as many levels as you wish, the large number of op-amps required for even a modest resolution makes the system very expensive. You only use a flash converter if you need a small conversion time.

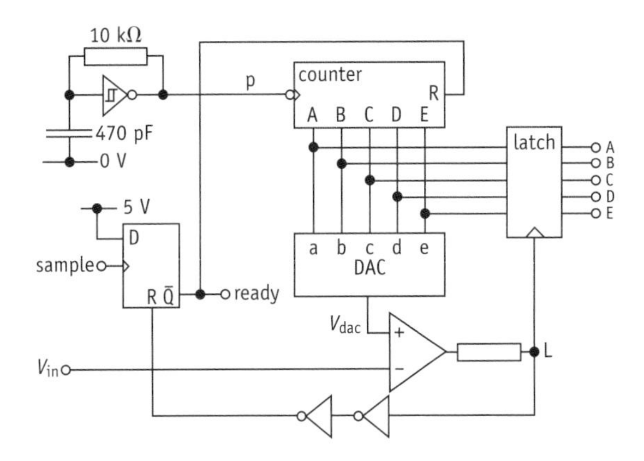

Fig 8.6 A 5-bit slope converter

Slope conversion

The **slope converter** of Fig 8.6 performs the ADC operation as follows.

- Reset the binary code to 00000 and wait until **sample** goes high.
- Run the binary code through a DAC to find its equivalent analogue signal V_{dac}.
- Use an op-amp to compare it with the input signal V_{in}. If it is greater, then latch the binary code onto the output bus and push **ready** high. Otherwise, add 00001 to the binary code and go back one step.

This procedure is shown in the timing diagram in Fig 8.7.

Conversion time

How long does it take for the ADC to make its measurement of the value of V_{in}? It clearly depends on its value, so let's consider the worst case. This is when $2^5 - 1 = 31$ falling edges at P are required before L goes high, the code is latched onto the output bus and the flip-flop is reset. In order to work out how long this takes, you need to find the period of the relaxation oscillator.

$T = ?$ $\quad$ $T = 0.5RC = 0.5 \times 10{\times}10^3 \times 470{\times}10^{-12} = 2.4{\times}10^{-6}$ s

$R = 10$ kΩ

$C = 470$ pF

So the maximum conversion time will be $31 \times 2.4{\times}10^{-6} = 7.3{\times}10^{-5}$ s or 73 μs.

Sample rate

It is important to match the conversion time to the sample rate. Suppose that the signal at V_{in} is a voice message in the range of 300 Hz to 3.4 kHz. It will have to be sampled at least twice in one cycle for its frequency to be correctly recreated by the DAC in the receiver. So the sample rate needs to be $2 \times 3\,400 = 6\,800$ Hz. The period of this needs to be greater than the conversion time.

$T = ?$

$f = 6\,800$ Hz

$$T = \frac{1}{f} = \frac{1}{6800} = 1.5 \times 10^{-4}\text{s or } 150\ \mu\text{s}$$

This is comfortably longer than the maximum conversion time of 73 μs.

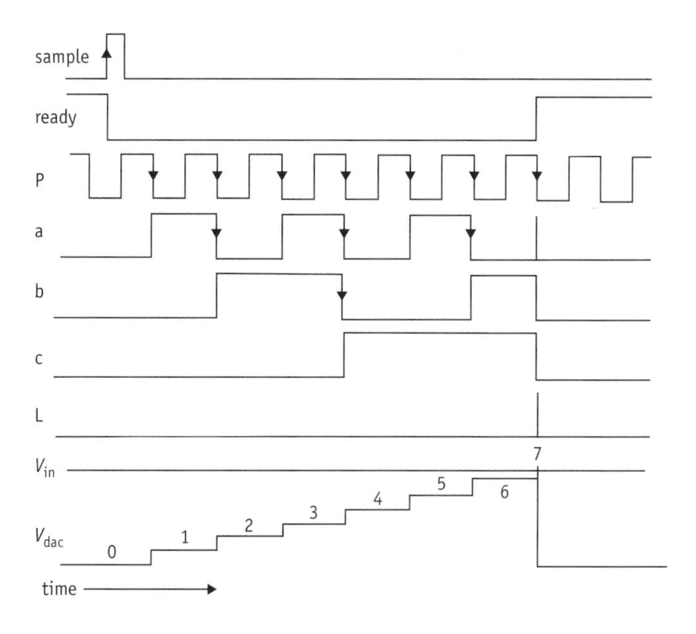

Fig 8.7 Timing diagram for the slope converter

DAC

The DAC in the ADC of Fig 8.6 is shown in Fig 8.8. It determines the range and resolution of the ADC. You find the **resolution** by calculating the value of V_{dac} when edcba = 00001. Since there is only current in one input resistor, you can ignore all the others and treat the left-hand op-amp as an inverting amplifier.

$V_{out} = ?$

$V_{in} = 5$ V

$R_{in} = 160$ kΩ

$R_f = 8$ kΩ

$$\frac{V_{out}}{V_{in}} = -\frac{R_f}{R_{in}}$$

$$V_{out} = -V_{in} \times \frac{R_f}{R_{in}} = -5 \times \frac{8 \times 10^3}{160 \times 10^3} = -0.25 \text{ V}$$

The right-hand amplifier has a gain of −1, so the resolution (the size of one step in the timing diagram) is 0.25 V. To find the **range**, you need to consider the number of inputs to the DAC – five. This gives $2^5 = 32$ different levels, starting at 0.00 V and ending at 31 × 0.25 = 7.75 V.

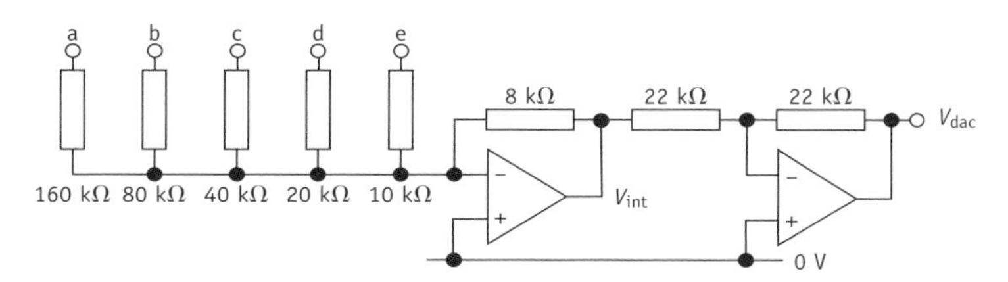

Fig 8.8 The DAC in the circuit of Fig 8.6

Compromise

The ideal ADC has a low conversion time, a large range and a good resolution. Unfortunately, since there is always a maximum speed at which oscillators and counters will work reliably, you often have to accept a poorer resolution to get the required conversion time. For example, think about what happens if you decide to improve the resolution by doubling the number of levels for the DAC. If the clock speed remains fixed, this will double the conversion time, halving the maximum input frequency that the ADC can reliably encode.

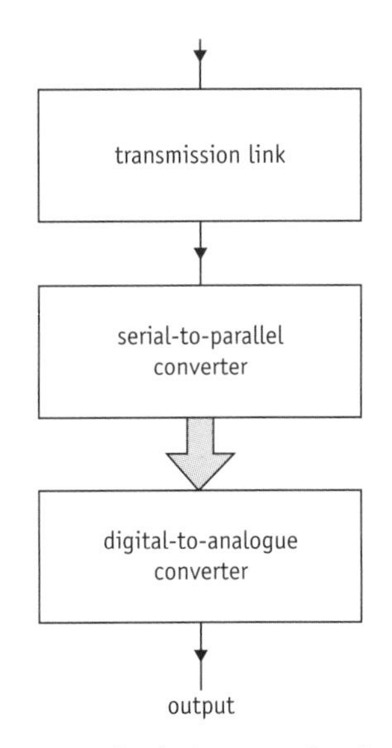

Fig 8.9 Block diagram for the receiver

Serial-to-parallel conversion

The block diagram of a digital transmission system receiver is shown in Fig 8.9.

You will recall that the **serial-to-parallel converter** (or **SPC**) takes bits one at a time as they arrive from the transmission link and places them as a binary word on a bus. A circuit which can do this is shown in Fig 8.10.

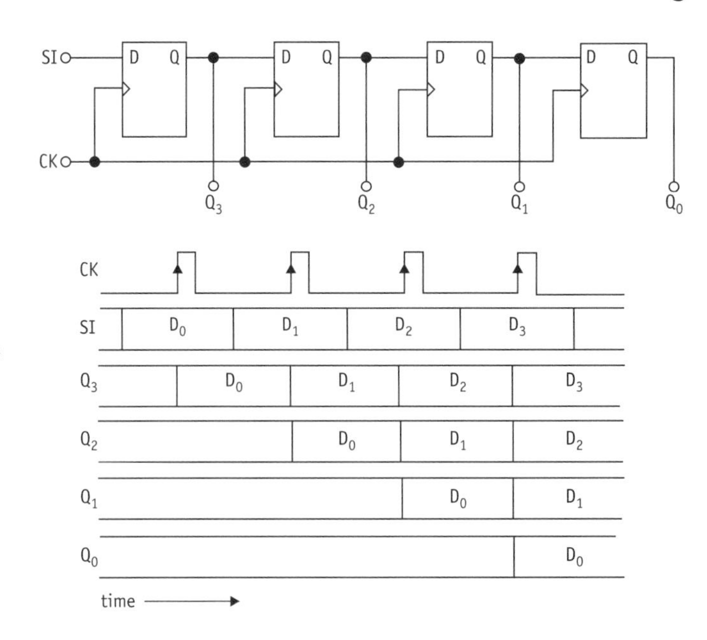

Fig 8.10 A 4-bit SIPO shift register

The sub-system is known as a **serial-in parallel-out** (or **SIPO**) shift register. It has a number of inputs and outputs.

- The serial input SI accepts bits as they arrive along the transmission link.
- Rising edges at the CK input cause the signal at the input of each D flip-flop to be copied to its output. As you can see from the timing diagram, each pulse at CK shuffles each bit stored in the register one place to the right.
- The word $Q_3Q_2Q_1Q_0$ at the output displays in parallel form the state of SI at the last of the four rising edges at CK.

Of course, for this to work, the clock pulses at CK need to be carefully timed to coincide with the arrival of bits at SI. This means that as well as carrying data, the transmission link also has to carry some timing information. Two ways of multiplexing this information will be explained later in this chapter.

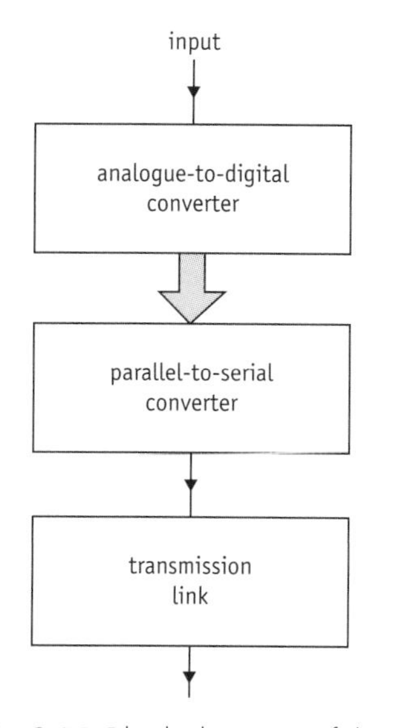

Fig 8.11 Block diagram of the transmitter

Parallel-to-serial conversion

The rate at which bits are sent down the transmission link is determined by the **parallel-to-serial converter** (or **PSC**) in the transmitter of Fig 8.11.

A circuit which performs the PSC function is shown in Fig 8.12.

It is called a **parallel-in serial-out** (or **PISO**) shift register. The timing diagram shows how it goes about its business.

- The PL input is raised high. This makes the multiplexers connect each bit of the parallel input word $D_3D_2D_1D_0$ to the data input of a different flip-flop.
- A rising edge at the CK input triggers the flip-flops into copying the bit at their data input to the output.
- The PL input is returned low, so that the system can behave like a SIPO shift register.
- The next three rising edges at CK shuffle the bits from left to right, so that D_0, D_1, D_2 and D_3 appear at the output SO one after the other.

Notice how SO goes high on the fifth clock pulse, as the logic 1 at the far left makes its way through the register to the output. As you will find out later, this important feature of the serial transmission systems allows the receiver to tell when the next word arrives.

Fig 8.12 A 4-bit PISO register

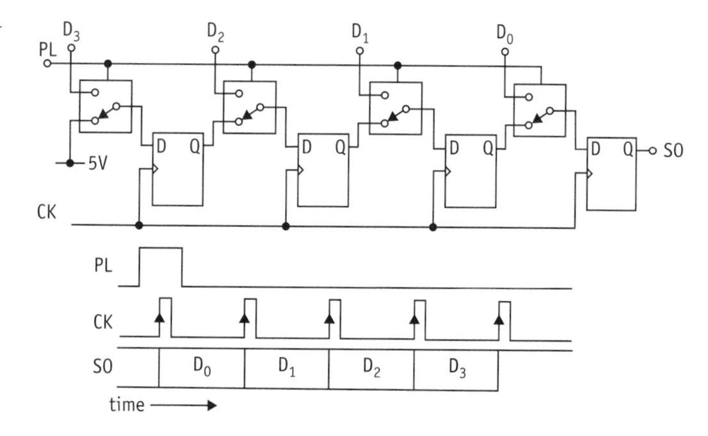

8.2 Synchronous transmission

Telephone networks routinely use **time-division multiplexing** (or **TDM**) to cram 32 different voice messages simultaneously down a single long-distance cable. The principle is quite simple.

- Divide each second into 8 000 different **frames**, each lasting for only 125 μs.
- Further divide each frame into 32 different **channels**, each lasting for 3.91 μs.

Information for each message is allocated a separate channel from all the other messages. So each message passes along the cable 8 000 times a second, in bursts of 3.91 μs. Since the moment-by-moment voltage of the message is coded as an 8-bit word, this means that each bit is placed on the cable for only 490 ns. The result is a stream of 2 048 000 bits emerging at the far end of the cable in each second. These bits have to be assembled into words, and the words from the same channel have to be fed one after the other to a DAC to recreate the original voice message. The only problem is, how can you tell which bit goes with which channel? The answer is to have accurate clocks and send some empty channels.

Multiplex, demultiplex

Figure 8.13 shows a simple four-channel TDM system, with four different messages (W, X, Y and Z) taking it in turns to pass down a single cable. The multiplexer on the left directs one of the four signals at its input onto the cable, depending on the output of a 2-bit counter. The demultiplexer on the right feeds the signals on the cable to one of four destinations, again depending on the state of a 2-bit counter.

Fig 8.13 Time-division multiplexing analogue signals

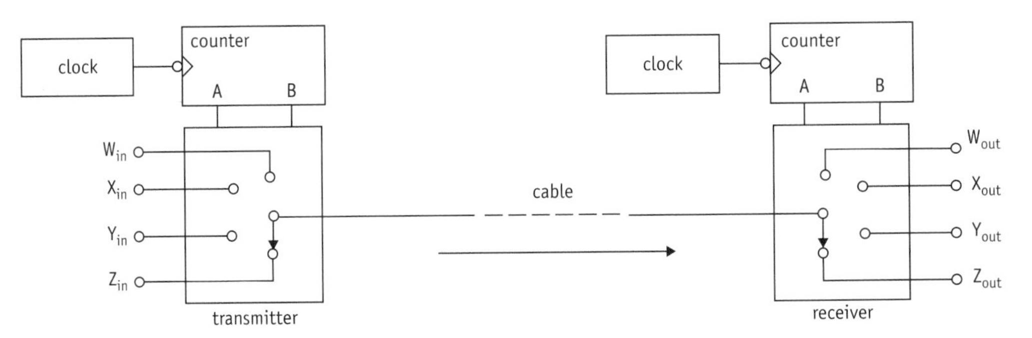

Consider the clock in the transmitter. Each falling edge at its output moves the counter onto the next state, changing the multiplexer input connected to the cable, as shown in the timing diagram of Fig 8.14. Both the multiplexer and demultiplexer have to rotate through their states in step with each other. This means that the clock in the receiver has to be **synchronised** with the transmitter clock.

Fig 8.14 Timing diagram for the transmitter

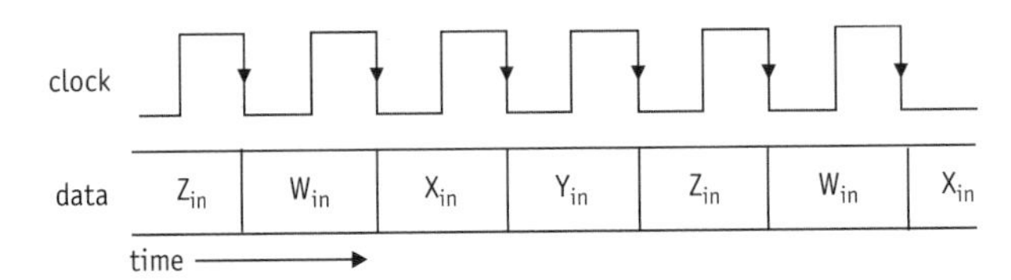

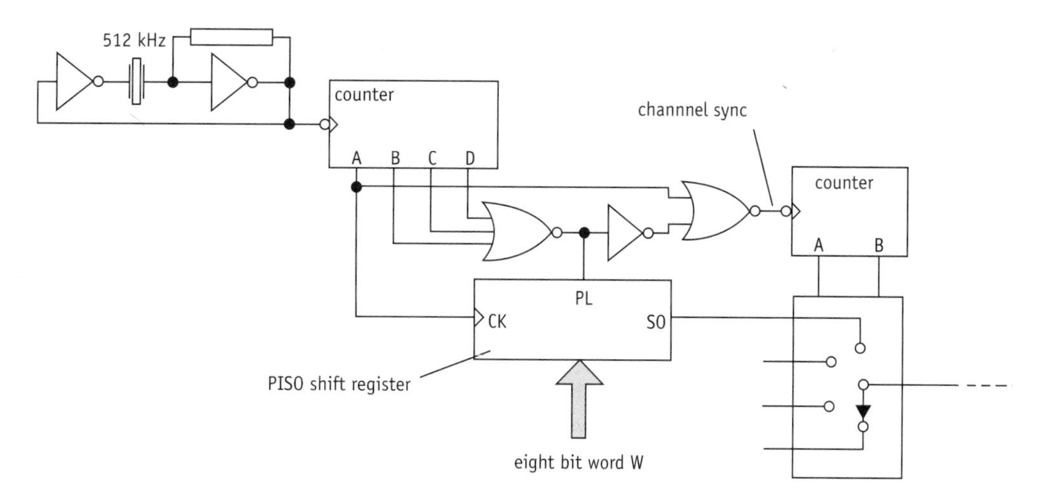

Fig 8.15 Partial circuit diagram for the synchronous transmitter

Transmitter

A partial circuit diagram for a **synchronous transmitter** is shown in Fig 8.15. It contains several sub-systems.

- The crystal oscillator produces a stable stream of pulses at a fixed rate of 512 kHz.
- The 4-bit counter uses the 512 kHz signal to generate signals for CK and PL, as shown in Fig 8.16.
- The PISO shift register feeds the word at its input (W in this case) one bit after the other to the cable via the multiplexer.
- The 2-bit counter changes the setting of the multiplexer on each falling edge at **channel sync**.

Notice how all the important timing signals (channel sync, CK and PL) have been obtained from one source of pulses (the 512 kHz oscillator) via a counter and some logic gates. This keeps all parts of the system carefully synchronised.

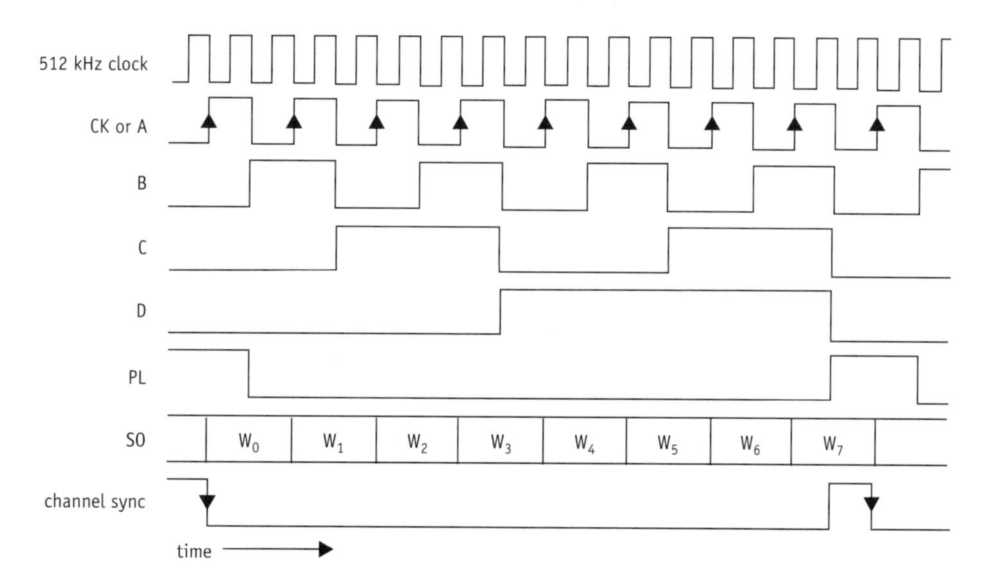

Fig 8.16 Timing diagram for a synchronous transmitter

Receiver

A synchronous receiver has to extract three separate pieces of information from the stream of bits coming down the cable towards it.

- When to start a new frame.
- When to change to the next channel.
- When to read each bit of a channel from the cable.

Two of these challenges are illustrated in the timing diagram of Fig 8.17. The channel sync signal needs to go low just as the first bit W_0 of the word arrives. Rising edges of the SIPO clock need to happen at the middle of the time allocated to each bit, so that the bits can be safely grabbed by a shift register and assembled into a binary word.

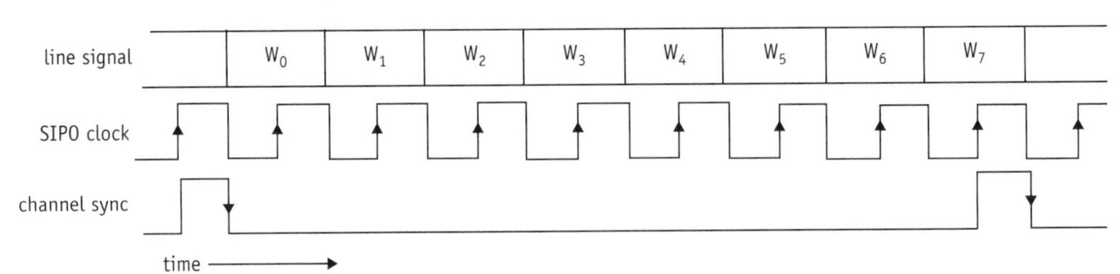

Fig 8.17 Timing diagram for one channel arriving at the receiver

The circuit of Fig 8.18 accomplishes these tasks. The timing diagram of Fig 8.19 on the next page shows how it goes about its business.

- Clock pulses from the oscillator enter the 6-bit counter at intervals of 0.5 μs.
- The C output changes state every 4 μs, triggering the SIPO shift register into grabbing a bit from the **line signal** passed on by the demultiplexer.
- The AND gate output goes low at the end of every eighth bit arriving on the line signal. This **channel sync** signal goes low at intervals of 32 μs, triggering the 2-bit counter into changing the state of the demultiplexer.

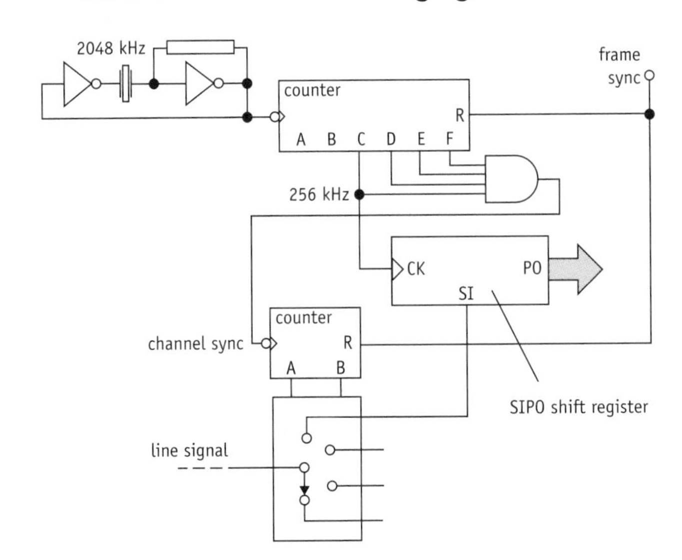

Fig 8.18 Partial circuit diagram for a synchronous receiver

All this supposes that the **frame sync** signal is low, so that both counters are free to react to pulses at their clock inputs. Another part of the receiver circuit generates the frame sync signal, using cues from the line signal as to when each frame starts.

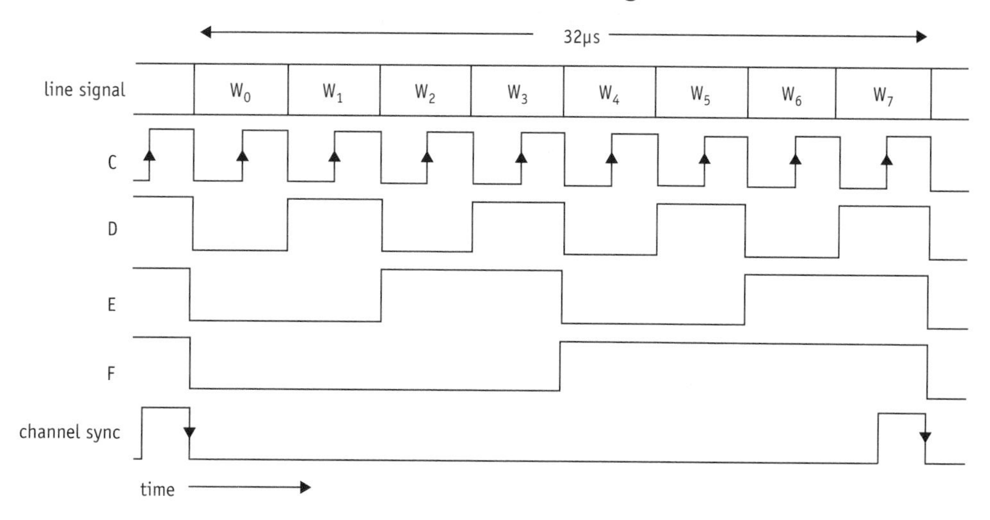

Fig 8.19 Timing diagram for the synchronous receiver

Frame sync

Synchronous transmission systems have to somehow encode the start of each new frame into the line signal. The system under consideration does it by imposing two rules on the contents of the four channels.

- One channel is left empty. So the word in Z is always 00000000.
- The most significant bit (or msb) of the other three channels must be a 1. So voltages are encoded as binary words between 10000000 and 11111111. The msb is, of course the last one to be transmitted within a channel.

So the task of the **frame sync generator** is to produce a brief pulse whenever the line signal has been low for 32 μs, as shown in the timing diagram of Fig 8.20. While that pulse is high, both of the counters in the synchronous counter of Fig 8.18 are reset. The demultiplexer connects the line signal to the W processing sub-system, and the 6-bit counter is unable to react to clock pulses from the crystal oscillator. Only when the frame sync signal goes low again does the system start latching bits from the line signal into the shift register.

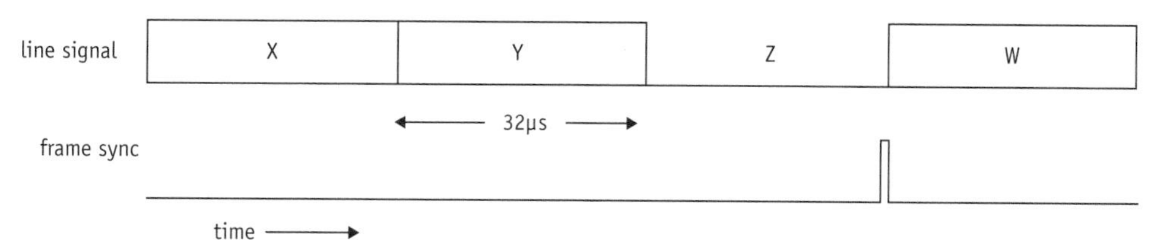

Fig 8.20 The frame sync pulses high at the end of the empty channel

Frame sync generator

The frame sync generator is shown in Fig 8.21. It uses the same crystal oscillator as the rest of the synchronous receiver, but the counter and the AND gate are repeat extras. The line signal resets the counter each time a 1 arrives during the transmission of the W, X or Y channels. This is going to happen at least once in every channel except for Z, which contains only 0. So consider what happens as the last bit of the Y channel enters the system.

- The last bit of the Y channel is always high, resetting the counter.
- The line signal then stays low for the entire Z channel, allowing the counter to react to the 0.5 μs pulses from the crystal oscillator.
- After 30 μs, 60 pulses will have been counted. This can only happen if all eight bits of the channel are 0.
- The output of the AND gate ($Q = F.E.D.C$) goes high for the next four pulses. This is the frame sync signal for the rest of the receiver, because it goes low right at the end of the Z channel.

The frame sync signal is unlikely to go low at exactly the same moment as a new frame arrives from the transmitter. This is because the timing can only be as precise as the intervals between pulses from the crystal oscillator. This is 0.5 μs, considerably less than the 4 μs for which each bit is on the line.

Fig 8.21 A frame sync generator

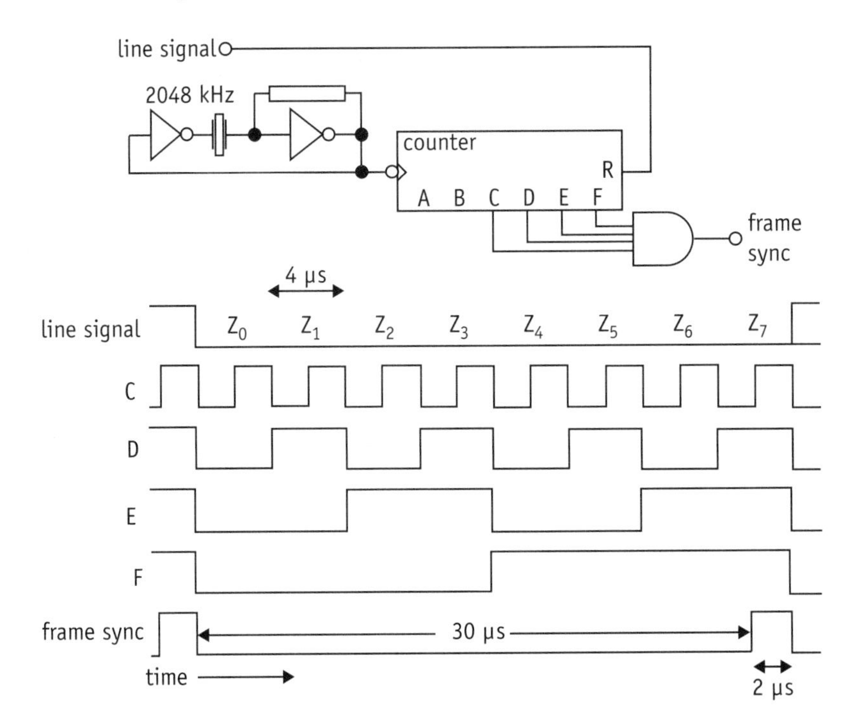

Bandwidth

The cable of a long-distance transmission system will dominate its cost, so it pays to multiplex as many different signals onto each line as possible. However, this increases the bit rate, requiring a greater bandwidth and therefore incurring greater attenuation for the high frequency signals. So there is always a trade-off between the number of channels, the frame rate, the word length and the range. Consider the standard TDM system used by telephone systems.

- Each frame contains 32 channels. Some of these will have to contain timing information to keep the receiver synchronised with the transmitter.
- Each channel carries an 8-bit word. This allows $2^8 = 256$ different levels for the moment-by-moment voltage of the voice message.
- There are 8 000 frames a second. Since each voice message must be sampled at least twice in each cycle, this places an upper limit of 4 kHz on the frequency of the voice signal, just above the standard 300 Hz – 3.4 kHz range for voice communication.

The bit rate for the cable is calculated as follows.

$$\text{bit rate} = \text{word length} \times \text{channel number} \times \text{frame rate}$$

$$\text{bit rate} = 8 \times 32 \times 8\,000 = 2\,048\,000\ \text{s}^{-1}$$

The bandwidth is the highest frequency that needs to be transmitted to get the information down the line. The signal which is going to generate the highest fundamental frequency will be one which alternates high with low, as shown in Fig 8.22.

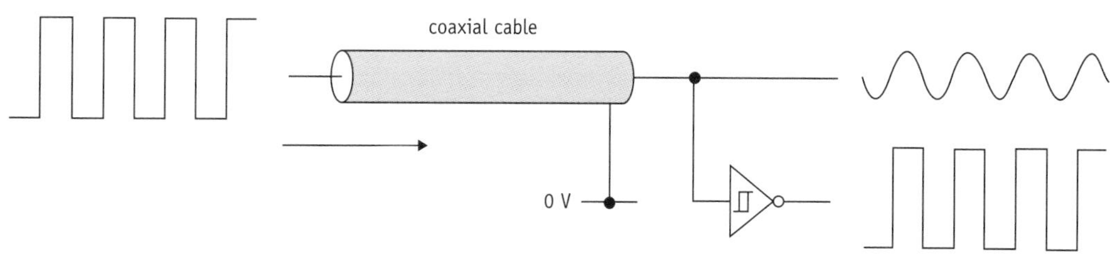

Fig 8.22 You only need to transmit the fundamental frequency

It won't happen all the time, but you still need to make sure that this **worst-case signal** will get down the line. Since you need two bits for every cycle of the fundamental frequency, the bit rate and bandwidth are connected by this formula.

$$\text{bandwidth} = \frac{\text{bit rate}}{2}$$

So a bit rate of $2.0 \times 10^6\ \text{s}^{-1}$ requires a bandwidth of 1.0×10^6 Hz. Standard twisted-pair cable has an attenuation of 0.1 per kilometre, somewhat high for long-distance communication. Coaxial cable, with one of the conductors running along the axis of the cylinder formed by the other, takes up more room but has a better attenuation, perhaps 0.5 per kilometre at 1 MHz, so does a better job. In any case, provided that the amplitude of the signal does not get too low, it can always be restored by a Schmitt trigger.

8.3 Asynchronous transmission

Not everyone knows that USB stands for **universal serial bus**. However, few users of ICT can be unaware that a USB cable uses just four wires to allow a computer high-speed two-way exchange of data with a variety of peripherals, such as cameras, printers, memories and mice. In fact, two of the wires are power lines, carrying +5 V and 0 V from the computer to the peripheral, leaving the other two as a shielded twisted-pair with a bandwidth of 6 MHz to carry the data. Digital data is sent back and forth along this cable as **packets** of bits that are transmitted **asynchronously**. This means that the arrival of a packet at its destination is unpredictable; it can arrive at any time. Circuits designed to receive **serial data** of this type have to do three things.

- Recognise the arrival of a packet.
- Read the bits at the correct time as they arrive.
- Check that all of the data has arrived successfully.

This section of the chapter will show you how this is done for devices using an RS-232 connection, the precursor of the USB port. Not only is RS-232 simpler than USB, it is similar to the protocol followed by microcontrollers when they use serial transfer for data.

RS-232

The table shows the voltage levels on the cable allocated to 1 and 0 by the RS-232 protocol.

logic level	line voltage
1	−12 V to −3 V
0	+12 V to +3 V

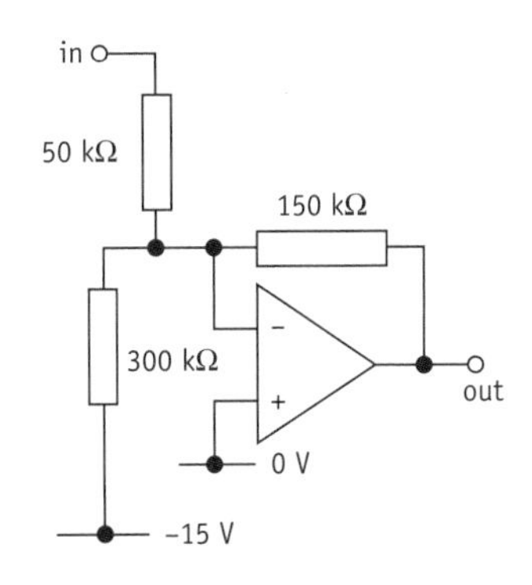

Fig 8.23 Logic level converter for an RS-232 transmitter

These **line voltage** ranges are not the same as those for logic gates (+5 V and 0 V). The summing amplifier shown in Fig 8.23 converts signals from logic gates into RS-232 signals.

Consider the line voltage at the output when the input is high.

$V_{out} = ?$

$R_f = 150\ \text{k}\Omega$

$V_1 = +5\ \text{V}$

$R_1 = 50\ \text{k}\Omega$

$V_2 = -15\ \text{V}$

$R_2 = 300\ \text{k}\Omega$

$$-\frac{V_{out}}{R_f} = \frac{V_1}{R_1} + \frac{V_2}{R_2}$$

$$V_{out} = -R_f\left(\frac{V_1}{R_1} + \frac{V_2}{R_2}\right) = -150 \times \left(\frac{5}{50} - \frac{15}{300}\right) = -7.5\text{V}$$

This in the middle of the range of −12 V to −3 V required for logic 1 when it is in transmission. You can verify for yourself that 0 V at the input raises the line voltage at the output to +7.5 V.

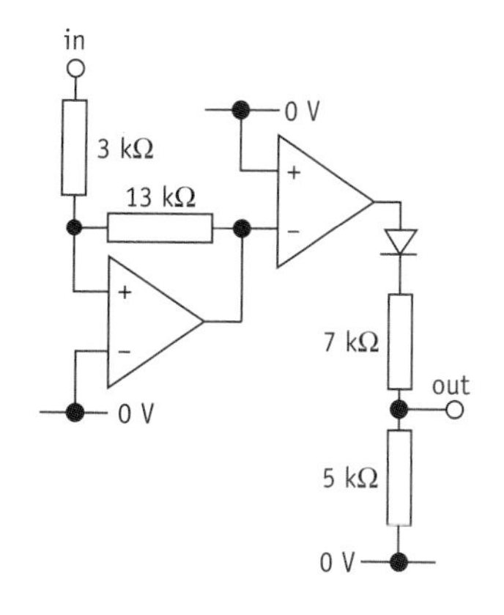

Fig 8.24 Logic level converter for an RS-232 receiver

Upon arrival at the receiver, another op-amp circuit, such as the one shown in Fig 8.24, is required to restore the voltage levels.

The circuit has two sub-systems.

- A Schmitt trigger with trip points at +3 V and – 3 V.
- A comparator and voltage divider to convert +13 V and −13 V to 0 V and +5 V.

Notice how the diode becomes reverse biased when the op-amp comparator output is saturated at −13 V, leaving the 5 kΩ resistor free to pull the output down to 0 V. Since RS-232 connections operate at relatively low bit rates (the maximum is 115 kbits s^{-1}), you don't have to worry about using high speed op-amps.

Start, stop

The timing diagram of Fig 8.25 shows the sequence of bits placed on the line when the transmitter sends out an 8-bit word in serial form. Three pieces of information are required.

- A **start** bit of 0. This will be used by the receiver to establish that a word is about to arrive.
- The eight bits of **data**, with the lsb first and the msb last.
- A **stop** bit of 1. This stays on the line, allowing the receiver to detect the next start bit. It can also provide a check at the receiver that the word has been correctly received – if the tenth bit is 0 then something has definitely gone wrong!

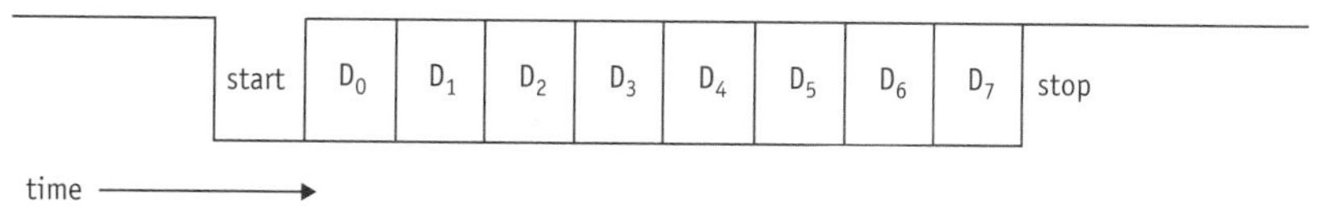

Fig 8.25 Timing diagram for serial transfer of a byte

Ten bits are required altogether. Each has to be on the line for the same length of time, depending on the bit rate. Let's assume the fastest rate of 115 000 bits s^{-1}, a typical speed for a computer communicating with a modem. You can calculate the time per bit as follows.

$$\text{bit time} = \frac{1}{\text{bit rate}} = \frac{1}{115000} = 8.7 \times 10^{-6}\ \text{s or } 9\ \mu\text{s}$$

The receiver therefore needs to latch bits from the line at intervals of 9 μs once the start bit has been recognised. Since ten bits are required for the transmission of each byte, the maximum data transfer rate is going to be a modest 11 500 bytes s^{-1}.

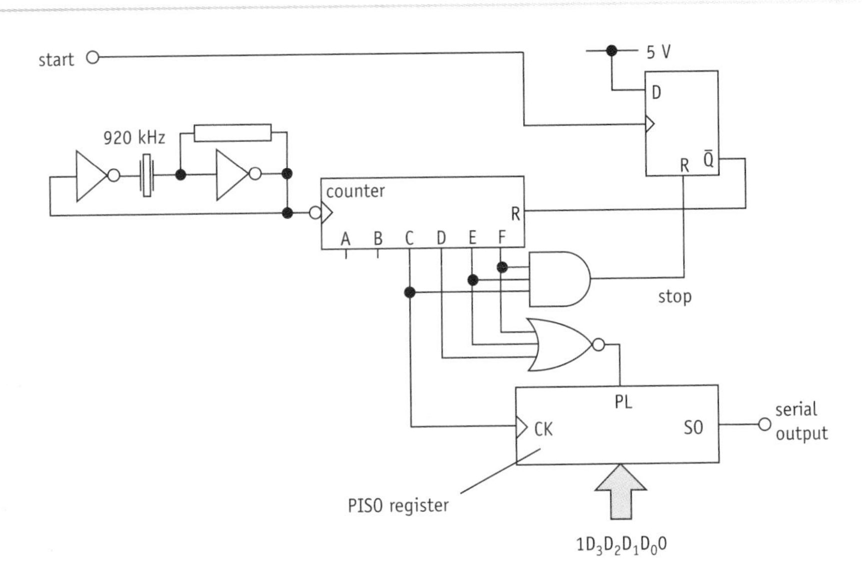

Fig 8.26 Serial transmitter of 4-bit words

Serial transmitter

A serial transmitter has to be triggered into action by some external device, usually a computer. Once it has been triggered, it needs to perform the following tasks.

- Load the start bit, data and stop bit into a PISO shift register.
- Apply clock pulses to the register so that the bits appear at the output one after the other at the correct rate.
- Leave the output high, and wait for the next triggering signal.

A circuit which does this for 4-bit words at a rate of 115 kbits s^{-1} is shown in Fig 8.26 above. The timing diagram of Fig 8.27 should help you to understand how it operates. Notice the role of the D flip-flop. A rising edge at the **start** terminal forces $\overline{Q}$ low, allowing the counter to respond to pulses from the 920 kHz crystal oscillator. Once all six bits have been placed on the line, the AND gate generates a brief pulse at **stop**, resetting the D flip-flop so that the counter is also reset and the system remains dormant until the next rising edge at **start**.

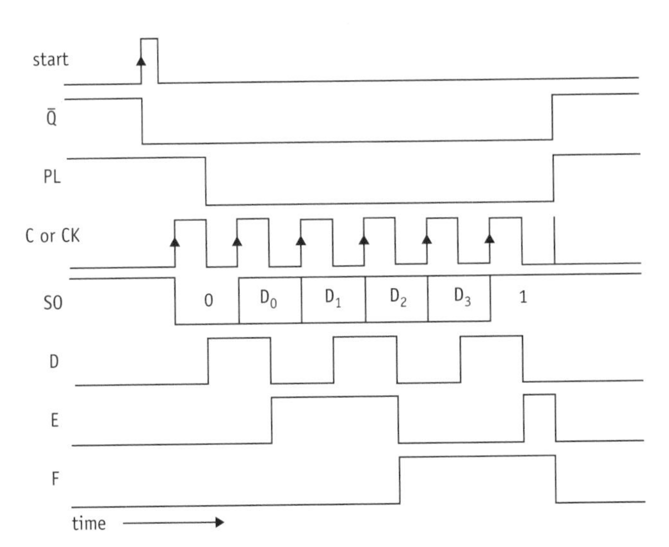

Fig 8.27 Timing diagram for the serial transmitter

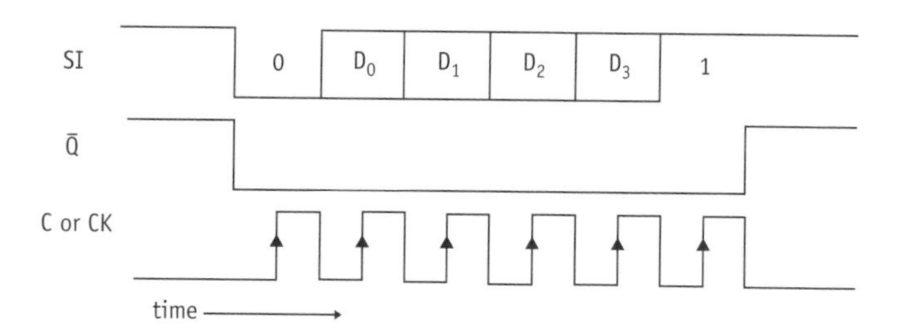

Fig 8.28 Timing diagram for the serial receiver

Serial receiver

Take a look at the timing diagram of Fig 8.28 above. It shows the tasks to be performed by a serial receiver which matches the transmitter on the previous page.

- The falling edge at the start of the word on the line has to set a D flip-flop, forcing $\overline{Q}$ low.
- The D flip-flop has to enable a series of rising edges at the clock input CK of a SIPO shift register. There must be six of them, at intervals of 9 μs.
- When all six bits have been safely latched from the line, the flip-flop must be reset. This leaves the word at the parallel output PO of the shift register.

A circuit which can do all this is shown in Fig 8.29. Notice that it uses the same crystal oscillator as the transmitter of Fig 8.26, so that once the receiver has been triggered by the start bit it can grab bits from the line at the right time. The period of both oscillators is a brief 1.1 μs.

$T = ?$

$f = 920$ kHz

$$T = \frac{1}{f} = \frac{1}{920 \times 10^3} = 1.09 \times 10^{-6}\text{s or } 1.1\ \mu\text{s}$$

One cycle of the signal at A will therefore last 2.2 μs, at B it will last for 4.4 μs and at C it will last for 8.8 μs. This means that any uncertainty of the timing at CK will be only 1.1 μs, much less than the 8.8 μs for which each bit is placed on the line by the transmitter.

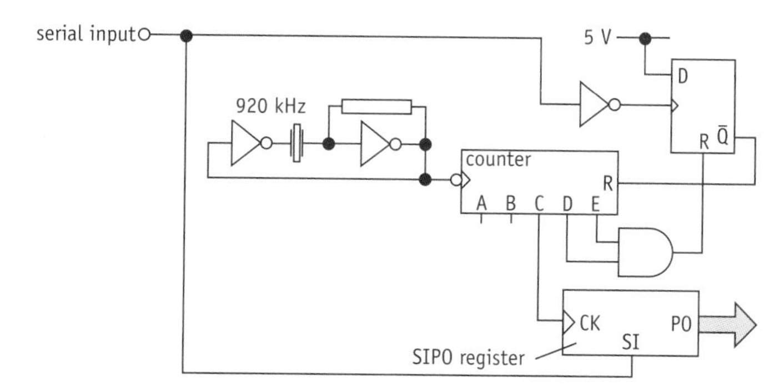

Fig 8.29 Serial receiver of 4-bit words

Parity check

The RS-232 protocal allows the inclusion of a **parity** bit between the data and stop bits. This can be used to verify that the data has been received correctly. The state of the parity bit is established in the transmitter by a series of EOR gates comparing the states of adjacent bits, as shown in Fig 8.30.

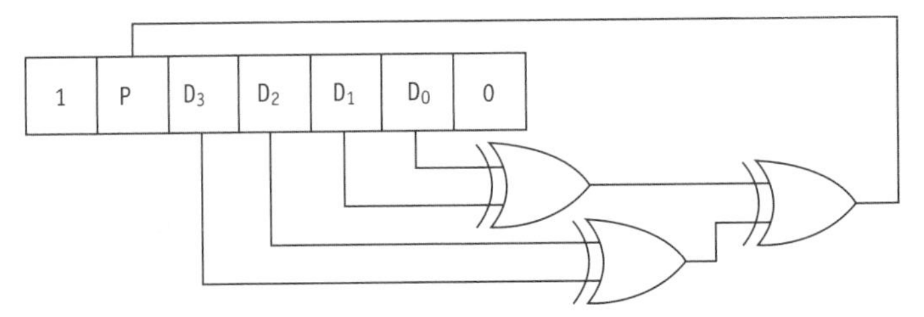

Fig 8.30 EOR gates determine the state of P in the transmitter

Suppose that $D_3D_2D_1D_0$ is 1011, with an odd number of bits which are 1. Then P is a 1, giving an even number of high bits in the whole word, ignoring the start and stop bits. You can verify for yourself, by trial-and-error, that the total number of high bits in the word $PD_3D_2D_1D_0$ will **always** be even. This should, of course, also be true at the receiver. Combining the four bits of the received word with EOR gates should result in a bit P which matches the received parity bit. If it doesn't, then something has gone wrong with the data transmission process and the received word can be rejected.

Ethernet

The protocols of RS-232 and its successor USB allow one device (the **host**) to exchange data in serial form with a number of different peripherals along one line. It works because the host makes all of the decisions about which device is allowed to access the line. The **ethernet** protocol goes one step further, allowing any number of devices (typically computers) to communicate with each other via a single line, without any one of them being in charge. The **internet** uses similar principles to allow any computer on the planet to communicate with almost any other. The situation is illustrated in Fig 8.31.

Each of the five devices can send or receive data in serial form. Only one output at a time can be connected to the line, so each device uses a tristate or analogue switch to disconnect its output when it isn't transmitting. On the other hand, inputs don't affect signals on the line, so they can always be connected to it. So all of the devices are listening to each other all the time, but they have to take it in turns to transmit data.

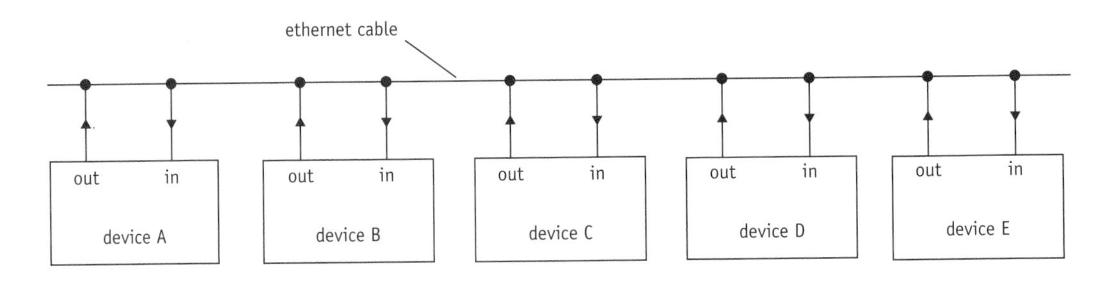

Fig 8.31 Five devices connected to the same serial cable

Packets

Each device places data on the line in the form of a **packet**. Each packet contains four different types of information.

- A **destination address**. Each device has its own unique address, so the destination address alerts one device connected to the line that it is being communicated with.
- A **source address**, the address of the device transmitting the packet.
- A **data payload**, the information being transmitted in binary form. This could be almost anything; text, music, video, software, etc.
- A **checksum**. At its simplest, this codes for the total sum of the bits which are high in the packet. This calculation can be performed by the receiver and the result compared with the checksum. If they don't agree, then something has gone wrong with the transmission process.

The timing diagram of Fig 8.32 shows the order in which the information is transmitted in the packet. The length of a typical packet is 1024 bits, of which 100 bits might be required for coding the destination address, source address and checksum. However, the typical bit rate is 10 Mbits s^{-1}, so it only takes 100 μs to transmit each packet.

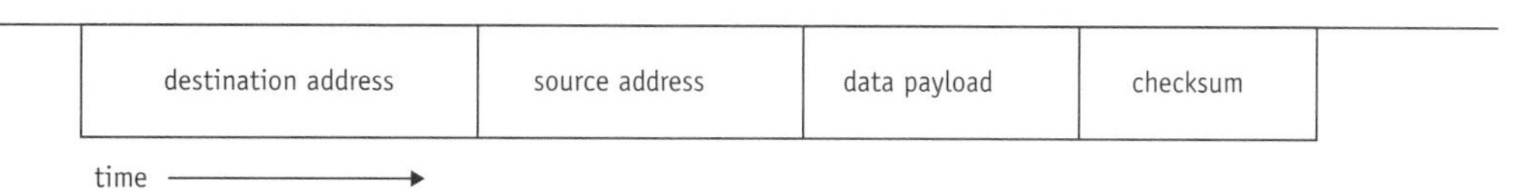

Fig 8.32 Timing diagram for a packet

Collision

Suppose that device A is required to send a packet to device C. This is the procedure it follows.

- Device A listens to the line until it has been quiet for at least 10 μs.
- It starts transmitting the packet, listening to the line at the same time.
- If the packet on the line doesn't match expectation, then some other device must also be trying to transmit. A **collision** has occured.
- Device A stops transmitting, and waits for a short random interval of time before returning to the first step.

Although this procedure means that collisions may be frequent, they won't be damaging. Both devices will stop transmitting immediately and try again at different times. Packets are therefore likely to arrive at their destination at haphazard intervals, depending on the number of devices which are trying to place packets on the line. Busy networks run slowly!

Questions

8.1 Digital transmission

1 **(a)** Draw a block diagram for a digital transmission system. Use the following blocks.

ADC DAC PSC SPC

(b) Describe the function of each block.

(c) Explain the advantages of transmitting information by digital signals instead of analogue signals.

(d) Explain the disadvantages of transmitting music by digital signals.

2 The MRB28 ADC has the following properties.

- a **conversion time** of 125 μs
- a **range** of 0 V to 1 V
- a **resolution** of 2 mV

(a) Explain the meaning of the term in **bold**.

(b) Show that the ADC produces a 9-bit word.

(c) By calculating the sample rate of the ADC, explain why it cannot be used to reliably encode analogue signals of frequency above 4 kHz.

3 A 2-bit flash converter has the characteristics given in the table.

voltage range / V	output word
0.00 to 0.49	11
0.50 to 0.99	10
1.00 to 1.49	00
1.50 to 1.99	01

(a) Show how four resistors in series can be used to generate reference voltages at 0.5 V, 1.00 V and 1.50 V from supply rails at +15 V and 0 V. Give resistor values and justify them.

(b) Show how op-amps and logic gates can be added to the resistors to make the flash converter.

(c) Explain how your system operates.

4 **(a)** Draw a block diagram for a slope conversion ADC. Use the following blocks.

clock comparator counter DAC flip-flop latch

(b) With the help of a timing diagram, explain how the system operates.

(c) The system can encode an analogue signal with 64 different levels at a rate of 6 800 samples per second. Use this information to calculate the clock speed and the number of bits for the counter.

(d) Explain the disadvantage of using a slope conversion ADC instead of a flash one.

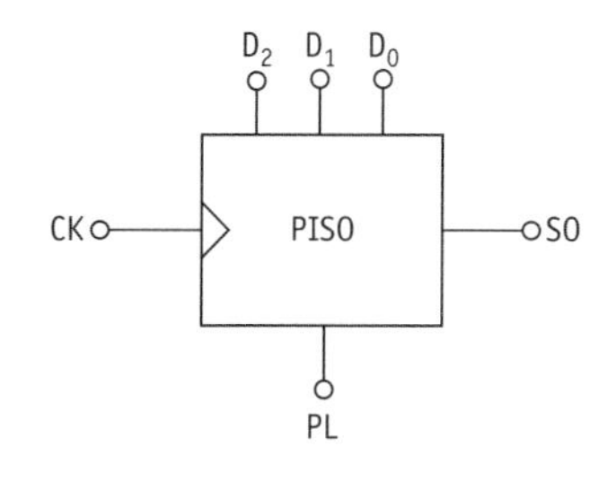

Fig 8.33 Question 5

5 This question is about the PISO shift register of Fig 8.33.

(a) Draw a circuit diagram for the register, using D flip-flops and multiplexers.

(b) With the help of a timing diagram, explain how the system operates.

(c) Draw a circuit diagram for a SIPO shift register which can convert the output of Fig 8.33 back into a 3-bit word.

8.2 Synchronous transmission

1 A telephone network uses time-division multiplexing to send 128 separate voice messages down a single optical fibre.

(a) Explain what is meant by the term **time-division multiplexing**.

(b) Each voice message is sampled 6 800 times each second, being coded as a 7-bit word. Calculate the bandwidth required for the optical fibre.

2 Two separate analogue signals are sent down the same link by time-division multiplexing.

(a) Draw a block diagram for the system. Use the following blocks.

ADC DAC clock counter demultiplexer link multiplexer PSC SPC

(b) With the help of a timing diagram, explain how the system operates.

(c) The signals cover the frequency range of 16 Hz to 16 kHz. Calculate a suitable frequency for the clocks.

(d) Explain why both clocks need to be in step. Describe how this can be done for the system.

8.3 Asynchronous transmission

1 Computers exchange packets of data along cables.

(a) Explain the function of the four pieces of information required in a packet.

(b) Explain how packets are placed on the cable one at a time.

2 In asynchronous transmission, words need to include **start**, **stop** and **parity** bits.

(a) Draw a timing diagram for the 3-bit word $D_2D_1D_0$ being sent down a line with start, stop and parity bits.

(b) Explain the function of the start and stop bits.

(c) Explain the function of the parity bit.

3 **(a)** Draw the block diagram for a serial transmitter. Use the following blocks.

clock counter D flip-flop logic system PISO register

(b) With the help of a timing diagram, explain how the system operates.

4 **(a)** Draw the block diagram for a serial receiver. Use the following blocks.

clock counter D flip-flop logic system SIPO register

(b) With the help of a timing diagram, explain how the system operates.

Appendices

Prefixes

prefix	multiplier
G	$\times 10^{9}$
M	$\times 10^{6}$
k	$\times 10^{3}$
m	$\times 10^{-3}$
μ	$\times 10^{-6}$
n	$\times 10^{-9}$
p	$\times 10^{-12}$

Boolean Algebra

$A.\overline{A} = 0$

$A + \overline{A} = 1$

$A.(B + C) = A.B + A.C$

$\overline{A.B} = \overline{A} + \overline{B}$

$\overline{A + B} = \overline{A}.\overline{B}$

$A + A.B = A$

$A.B + \overline{A}.C = A.B + \overline{A}.C + B.C$

Assembler codes

Code	Function
add Sd, Ss	Add the byte in Ss to the byte in Sd and store the result in Sd
sub Sd, Ss	Subtract the byte in Ss from the byte in Sd and store the result in Sd.
and Sd, Ss	Logical AND the byte in Ss with the byte in Sd and store the result in Sd.
eor Sd, Ss	Logical EOR the byte in Ss with the byte in Sd and store the result in Sd.
inc Sd	Add one to the byte in register Sd.
dec Sd	Subtract one from the byte in register Sd.
shl Sd	Shift the byte in Sd one bit to the left, putting a 0 into the lsb.
shr Sd	Shift the byte in Sd one bit to the right, putting a 0 into the msb.
jp e	Jump to label e.
jz e	Jump to label e if the result of the last of the above operations was zero.
jnz e	Jump to label e if the result of the last of the above operations was not zero.
movi Sd,n	Copy the byte n into register Sd.
mov Sd, Ss	Copy the byte from register Ss into Sd.
in Sd, I	Copy the byte at the input port to Sd.
out Q, Ss	Copy the byte in Ss to the output port.
rcall s	Jump to the subroutine at label s.
ret	Return to main program from subroutine.

Formulae

$R = \frac{V}{I}$	R is resistance in ohms (Ω) V is voltage drop in volts (V) I is current in amps (A)
$P = VI$	P is the power in watts (W) V is the voltage drop in volts (V) I is the current in amps (A)
$\tau = RC$	τ is the time constant in seconds (s) R is the resistance in ohms (Ω) C is the capacitance in farads (F)
$T = 0.7RC$	T is the pulse duration of a monostable in seconds (s) R is the resistance in ohms (Ω) C is the capacitance in farads (F)
$T = 0.5RC$	T is the period of a relaxation oscillator in seconds (s) R is the resistance in ohms (Ω) C is the capacitance in farads (F)
$f = \frac{1}{T}$	f is the frequency of a signal in hertz (Hz) T is the period of the signal in seconds (s)
$G = \frac{V_{out}}{V_{in}}$	G is the voltage gain of an amplifier V_{in} is the voltage of the input signal in volts (V) V_{out} is the voltage of the output signal in volts (V)
$V_{out} = A(V_+ - V_-)$	V_{out} is the output voltage of an op-amp in volts (V) A is the open-loop gain of the op-amp V_+ is the voltage at the non-inverting input in volts (V) V_- is the voltage at the inverting input in volts (V)
$G = 1 + \frac{R_f}{R_d}$	G is the voltage gain of a non-inverting amplifier R_f is the resistance of the feedback resistor in ohms (Ω) R_d is the resistance of the pulldown resistor in ohms (Ω)
$G = -\frac{R_f}{R_{in}}$	G is the voltage gain of an inverting amplifier R_f is the resistance of the feedback resistor in ohms (Ω) R_{in} is the resistance of the input resistor in ohms (Ω)

$-\frac{V_{out}}{R_f} = \frac{V_1}{R_1} + \frac{V_2}{R_2}\ldots$	V_{out} is the voltage at the output of a summing amplifier in volts (V) R_f is the resistance of the feedback resistor in ohms (Ω) V_n is the voltage at an input in volts (V) R_n is the resistance of an input resistor in ohms (Ω)
$f_0 = \frac{1}{2\pi RC}$	f_0 is the break frequency of a filter in hertz (Hz) R is the resistance of the resistor in ohms (Ω) C is the capacitance of the capacitor in farads (F)
$I_d = g_m(V_{gs} - V_{th})$	I_d is the drain current of a MOSFET in amps (A) g_m is the transconductance in siemens (S) V_{gs} is the gate-source voltage in volts (V) V_{th} is the threshold voltage in volts (V)
$G = -g_m R_d$	G is the voltage gain of a MOSFET amplifier g_m is the transconductance in siemens (S) R_d is the resistance of the drain resistor in ohms (Ω)
$\Delta V_{out} = -V_{in}\frac{\Delta t}{RC}$	ΔV_{out} is the change in output voltage of a ramp generator in volts (V) V_{in} is the input voltage in volts (V) R is the resistance of the resistor in ohms (Ω) C is the capacitance of the capacitor in farads (F)
$X_L = 2\pi fL$	X_L is the reactance of an inductor in ohms (Ω) f is the frequency of an a.c. current in hertz (Hz) L is the inductance of an inductor in henries (H)
$X_C = \frac{1}{2\pi fC}$	X_C is the reactance of a capacitor in ohms (Ω) f is the frequency of an a.c. current in hertz (Hz) C is the capacitance of a capacitor in farads (F)
$f_0 = \frac{1}{2\pi\sqrt{LC}}$	f_0 is the resonant frequency of a tuned circuit in hertz (Hz) L is the inductance of an inductor in henries (H) C is the capacitance of a capacitor in farads (F)

Circuit symbols

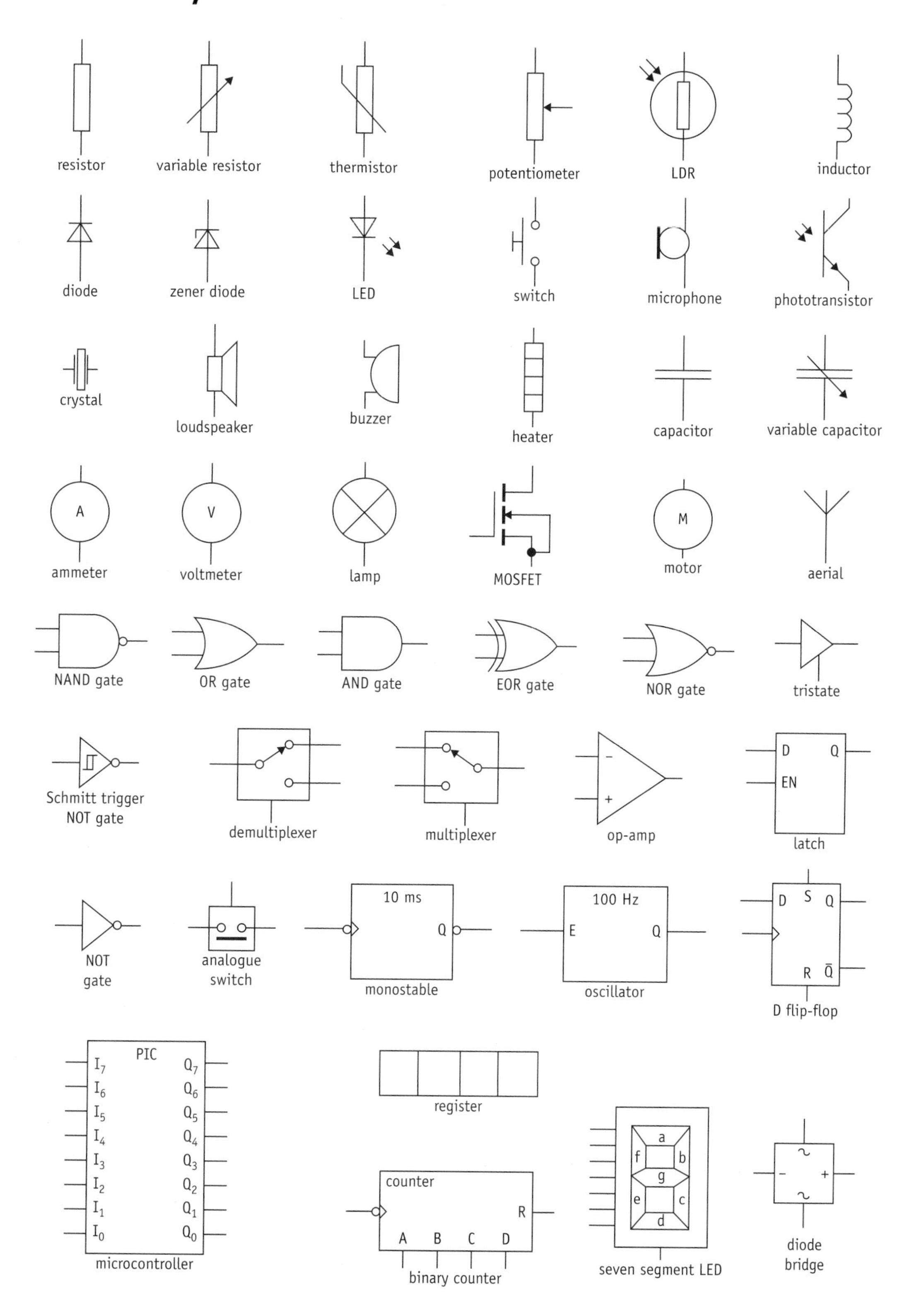

Index